SOLUTIONS MANUAL

PHYSICAL CHEMISTRY
FIFTH EDITION

ROBERT A. ALBERTY

John R Freeman
apt 408 Twin Towers
Please Return (Reward)
508

John Wiley & Sons, Inc.
New York Chichester Brisbane Toronto

ISBN 0 471 04749 X
Printed in the United States of America

10 9 8 7 6 5 4 3 2 1

PREFACE

This manual gives solutions for 80 percent of the problems in the first set of the text R.A. Alberty & F. Daniels, *Physical Chemistry*, 5th ed., Wiley, New York, 1979, and it gives answers for all of the problems in the second set.

Working problems is an important part of learning physical chemistry. Not all knowledge of physical chemistry is quantitative, but much of it is. Since physical chemistry utilizes physics and mathematics to predict and interpret chemical phenomena, there are many opportunities to use quantitative methods.

The availability of hand-held electronic calculators has made it much easier to work physical chemistry problems, but such a calculator is not necessary to work problems in this book. The units of physical quantities are always shown in solving problems in this manual. It is important to develop the habit of using units and cancelling them to obtain the units for the answer because this helps prevent errors.

In this book, as in your handwritten lecture notes, there is no distinction between italic (sloping) and roman (upright) type, but it is important to note that in the printed literature italic type is used for symbols for physical quantities and roman type is used for units.

This problem set has been built up over a long period of years (66 years if we go back to the first edition of *Outlines of Theoretical Chemistry* by Dr. Frederick H. Getman) and many individuals have contributed to it. Dr. Farrington Daniels developed many problems for physical chemistry instruction in the 45 years he was involved in *Outlines of Theoretical Chemistry* and *Physical Chemistry*.

Robert A. Alberty

Cambridge, Mass.
July 1978

PREFACE

[illegible]

CONTENTS

Part One: Thermodynamics

CHAPTER 1. First Law of Thermodynamics

1.1 How much work is done when a person weighing 75 Kg (165 lbs.) climbs the Washington monument, 555 ft. high? How many kilocalories must be supplied to do this muscular work, assuming that 25% of the energy produced by the oxidation of food in the body can be converted into muscular mechanical work?

SOLUTION:

$w = mgh$

work = (mass)(acceleration of gravity)(height)

$= (75\text{ kg})(9.806\text{ ms}^{-2})(555\text{ ft.}) \times (12\text{ in ft}^{-1})$

$\times (2.54 \times 10^{-2}\text{ min}^{-1})$

$= 1.244 \times 10^{5}\text{ J}$

The energy needed is four times greater than the work done.

$E = 4\,(1.244 \times 10^{5}\text{ J})/(4.184\text{ J cal}^{-1})(10^{3}\text{cal kcal}^{-1})$

$= 118.9\text{ kcal}$

1.2 A mole of liquid water is vaporized at 100°C and 760 torr. The heat of vaporization is 9725 cal mol^{-1}. What are the values of (a) w_{rev}, (b) q, (c) ΔU, and (d) ΔH?

SOLUTION:

(a) Assuming that water vapor is a perfect gas and that the volume of liquid water is negligible,

$$w = -P\Delta V = -RT$$

$$= -(1.987 \text{ cal K}^{-1} \text{ mol}^{-1})(373.15 \text{ K})$$

$$= -741.4 \text{ cal mol}^{-1}$$

(b) The heat of vaporization is 9725 cal mol^{-1}, and, since heat is absorbed, q has a positive sign.

$$q = 9725 \text{ cal mol}^{-1}$$

(c) $\Delta U = q + w$

$$= 9725 \text{ cal mol}^{-1} - 741 \text{ cal mol}^{-1}$$

$$= 8984 \text{ cal mol}^{-1}$$

(d) $\Delta H = \Delta U + \Delta(PV) = \Delta U + P\Delta V$

$$= \Delta U + RT$$

$$= 8984 \text{ cal mol}^{-1} + (1.987 \text{ cal K}^{-1} \text{ mol}^{-1})(373.15 \text{ K})$$

$$= 9725 \text{ cal mol}^{-1}$$

1.3 Considering H_2O to be a rigid nonlinear molecule, what value of C_P for the gas would be expected classically? If vibration is taken into account what value is expected? Compare these values of C_P with the actual values at 298 and 3000 K in Table A.1.

SOLUTION:

A rigid molecule has translational and rotational energy. The translational contribution to C_V is $\frac{3}{2}R = 2.981$ cal K^{-1} mol^{-1}.

Since H_2O is a nonlinear molecule, it has three rotational degrees of freedom, and so the rotational contribution to C_V is $\frac{3}{2}R = 2.981$ cal K^{-1} mol^{-1}.

Thus C_P for the rigid molecule is

$$C_P = C_V + R = 7.948 \text{ cal K}^{-1} \text{ mol}^{-1}$$

Since H_2O is a nonlinear molecule, the number of vibrational degrees of freedom is $3N - 6 = 3$. Since each vibrational degree of freedom contributes R to the heat capacity, classical theory predicts.

$$C_P = 7.948 \text{ cal K}^{-1} \text{ mol}^{-1} + 3R$$

$$= 13.909 \text{ cal K}^{-1} \text{ mol}^{-1}$$

The experimental value of C_P at 298K is 8.025 cal K^{-1} mol^{-1}, which is only slightly higher than the value expected for a rigid molecule. The experimental value of C_P at 3000K is 13.304 cal K^{-1} mol^{-1}, which is only slightly less than the classical expectation for a vibrating water molecule.

1.4 The equation for the molar heat capacity of n-butane is

$$C_P = 4.64 + 0.0558T$$

where C_P is given in cal K^{-1} mol^{-1}

Calculate the heat necessary to raise the temperature of 1 mole from 25 to 300°C at constant pressure.

SOLUTION:

$$q = \int_{T_1}^{T_2} C_P dT = 4.64(T_2 - T_1) + \tfrac{1}{2}(0.0558)(T_2^2 - T_1^2)$$

$$= 4.64\ (573.15 - 298.15) + \tfrac{1}{2}(0.0558)(573.15^2 - 298.15^2)$$

$$= 7961\ \text{cal mol}^{-1}$$

1.6 One mole of nitrogen at 25° C and 760 torr is expanded reversibly and isothermally to a pressure of torr (a) What is the value of w? What is the value of w if the temperature is 100° C?

SOLUTION:

(a) $w = -\int P dV = -RT \ln \frac{P_1}{P_2}$

$$= -(1.987\ \text{cal K}^{-1}\text{mol}^{-1})(298.15\text{K}) \ln \frac{760}{100}$$

$$= -1202\ \text{cal mol}^{-1}$$

(b) $w = (-1202\ \text{cal mol}^{-1})(373.15\ \text{K})/(298.15\ \text{K})$

$$= -1504\ \text{cal mol}^{-1}$$

1.7 A mole of argon is allowed to expand adiabatically from a pressure of atm and 298.15 K to 1 atm. What is the final temperature and how much work can be done?

SOLUTION:

$$\gamma = C_p/C_v = (\tfrac{5}{2}R)/(\tfrac{3}{2}R) = \tfrac{5}{3}$$

$$(\gamma - 1)/ = 2/5$$

$$\frac{T_1}{T_2} = \left(\frac{P_1}{P_2}\right)^{(\gamma-1)/\gamma}$$

$$T_2 = (298.15\text{K})(1/10)^{2/5} = 118.70\ \text{K}$$

$$w = \int_{T_1}^{T_2} C_V dT = \tfrac{3}{2}R(T_2 - T_1)$$

$$= \tfrac{3}{2}(1.987\text{cal K}^{-1}\text{mol}^{-1})(118.70\text{K} - 298.15\ \text{K})$$

$$= -535\ \text{cal mol}^{-1}$$

Thus the maximum work that can be done on the surroundings is 535 cal mol^{-1}.

1.8 A tank contains 20 liters of compressed nitrogen at 10 atm and 25° C. Calculate the maximum work (in calories) which can be obtained when gas is allowed to expand to 1 atm pressure (a) isothermally and (b) adiabatically.

SOLUTION:

(a) First we will do the calculation for one mole of perfect gas. For an isothermal expansion

$$W_{rev} = -RT \ln \frac{P_2}{P_1}$$

$$= -(1.987 \text{ cal K}^{-1} \text{ mol}^{-1})(298.15 \text{ K}) \ln 10$$

$$= -1364 \text{ cal mol}^{-1}$$

There are (10 atm)(20 L)/(0.08205 L atm K^{-1} mol^{-1})(298.15 K) = 8.18 moles. Therefore, the maximum work done on the surroundings for this much gas is 11.2 kcal mol^{-1}.

(b) For the adiabatic expansion we will assume that $\gamma = C_P/C_V$ has the value it has at room temperature. From Table A.2

$$\gamma = 6.961/(6.961 - 1.987)$$

$$= 1.399$$

$$\frac{T_1}{T_2} = \frac{P_1}{P_2}^{(\gamma-1)/\gamma}$$

$$T_2 = (298.15 \text{ K})(1/10)^{0.285} = 154.7 \text{ K}$$

$$w = \int_{T_1}^{T_2} C_V \, dT = C_V (T_2 - T_1)$$

$$= (4.974 \text{ cal K}^{-1} \text{mol}^{-1})(154.7 - 298.15 \text{ K})$$

$$= -714 \text{ cal mol}^{-1}$$

For 8.18 moles the maximum work done on the surroundings is 5.84 kcal mol^{-1}.

1.10 Calculate the second virial coefficient for hydrogen at 0°C from the fact that molar volumes at 50, 100, 200, and 300 atm are 0.4634, 0.2386, 0.1271, and 0.09004 L mol^{-1}, respectively.

SOLUTION:

$$\frac{PV}{RT} = 1 + \frac{B}{V} + \frac{C}{V^2} + \cdots$$

P/atm	50	100	200	300
V/L mol^{-1}	0.4634	0.2386	0.1271	0.09004
PV/RT	1.034	1.065	1.134	1.205
(1/V)/mol L^{-1}	2.158	4.191	7.868	11.106

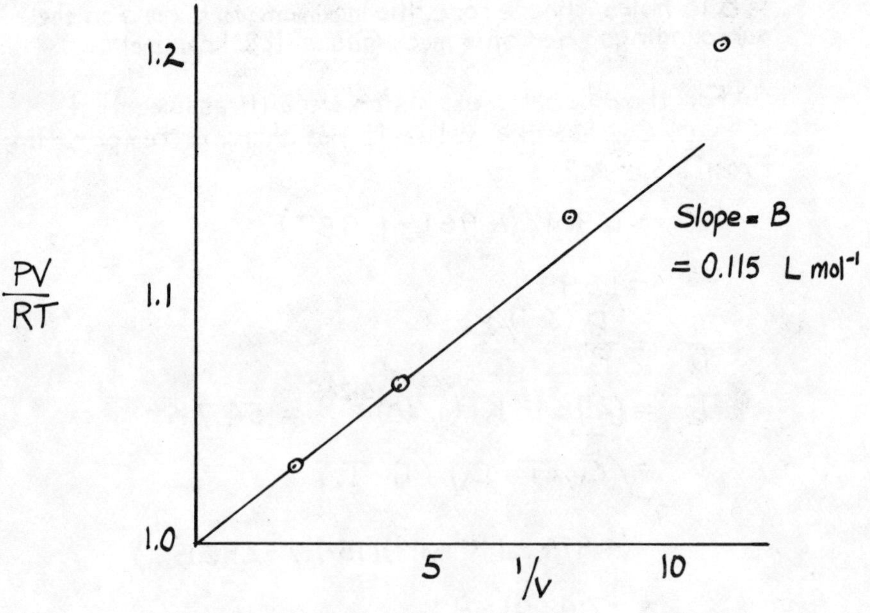

1.11 The second virial coefficient B of methyl isobutyl ketone is $-1580\ \text{cm}^3\ \text{mol}^{-1}$ at 120°C. Compare its compressibility factor at this temperature with that of a perfect gas at 1 atm.

SOLUTION:

$$Z = \frac{PV}{RT} = 1 + \frac{B}{V} = 1 + \frac{BP}{RT}$$

In the B/V term, V may be replaced by RT/P if $Z \approx 1$, since the approximation is made in a small correction term. At 1atm

$$Z = 1 + \frac{(-1.58\ \text{L mol}^{-1})(1\ \text{atm})}{(0.08205\ \text{L atm K}^{-1}\text{mol}^{-1})(393.15\ \text{K})}$$

$$= 0.951$$

The compressibility factor for a perfect gas is of course unity.

1.12 Show that for a gas of spherical molecules b in the van der Waals equation is four times the molecular volume times Avogadro's constant.

SOLUTION:

The molecular volume for a spherical molecule is

$$\frac{4}{3}\pi\left(\frac{d}{2}\right)^3 = \frac{\pi}{6}d^3$$

where d is the diameter. Since the center of a second spherical molecule cannot come within a distance d of the center of the first spherical molecule, the excluded volume per pair of molecules is

$$\frac{4}{3}\pi d^3$$

The constant b in van der Waals equation is the excluded volume per molecule times Avogadro's constant

$$b = \frac{2}{3}\pi d^3 N_A = 4\left(\frac{\pi}{6}d^3\right)N_A$$

1.13 In an adiabatic calorimeter, oxidation of 0.4362 gram of naphthalene caused a temperature rise of 1.707° C. The heat capacity of the calorimeter and water was 2460 cal K^{-1}. If corrections for oxidation of the wire and residual nitrogen are neglected, what is the enthalpy of combustion of naphthalene per mole?

SOLUTION:

$$\Delta H = \frac{(2460 \text{ cal K}^{-1})(1.707\text{K})(128.19\text{g mol}^{-1})}{(0.4362\text{g})(1000\text{cal Kcal}^{-1})}$$

$$= -1234 \text{ Kcal mol}^{-1}$$

The following reactions might be used to power rockets:

(1) $H_2(g) + \frac{1}{2}O_2(g) = H_2O(g)$

(2) $CH_3OH(l) + 1\frac{1}{2}O_2(g) = CO_2(g) + 2H_2O(g)$

(3) $H_2(g) + F_2(g) = 2HF(g)$

(a) Calculate the enthalpy changes at 25°C for each of these reactions per kilogram of reactants.

(b) Since the thrust is greater when the molar mass of the exhaust gas is lower, divide the heat per kilogram by the molar mass of the product (or the average mass in the case of reaction 2) and arrange the above reactions in order of effectiveness on the basis of thrust.

SOLUTION:

(a) (1) $\Delta H = -57.80 \text{ kcal mol}^{-1}$

$= (-57.80 \text{ kcal mol}^{-1})(1000 \text{ g kg}^{-1})/18 \text{ g mol}^{-1})$

$= -3211 \text{ kcal Kg}^{-1}$

(2) $\Delta H = -94.05 + 2(-57.80) + 57.04 = -52.61 \text{ kcal mol}^{-1}$

$= (-152.61 \text{ kcal mol}^{-1})(1000 \text{ g kg}^{-1})/(80 \text{g mol}^{-1})$

$= -1908 \text{ kcal Kg}^{-1}$

(3) $\Delta H = 2(-64.8) = -129.6 \text{ kcal mol}^{-1}$

$= (-129.6 \text{ kcal mol}^{-1})(1000 \text{ g kg}^{-1})/(40 \text{ g mol}^{-1})$

$= -3240 \text{ K cal Kg}^{-1}$

(b) (1) $-3211/18 = -178$

(2) $\dfrac{-1908}{\frac{1}{3}(44+2\times 18)} = 71.5$

(3) $-3240/20 = -162$

$\therefore$ (1) > (3) > (2)

1.16 Calculate the heat of hydration of Na_2SO_4 (s) from the integral heats of solution of Na_2SO_4 (s) and $Na_2SO_4 \cdot 10H_2O$(s) in infinite amounts of H_2O, which are -0.56 kcal mol^{-1} and 18.85 kcal mol^{-1}, respectively. Enthalpies of hydration cannot be measured directly because of the slowness of the phase transition.

SOLUTION:

$Na_2SO_4\,(s) = Na_2SO_4\,(aq)$	$\Delta H° = -0.56 \text{ kcal mol}^{-1}$
$Na_2SO_4\,(aq) = Na_2SO_4 \cdot 10H_2O$	$\Delta H° = -18.85 \text{ kcal mol}^{-1}$
$Na_2SO_4\,(s) + 10H_2O\,(\ell) = Na_2SO_4 \cdot 10H_2O(s)$	

1.17 Calculate the integral heat of solution of one mole of $HCl(g)$ in $200\ H_2O(l)$.

$$HCl(g) + 200H_2O(l) = HCl \text{ in } 200H_2O$$

SOLUTION:

$$\Delta H^\circ_{298} = \Delta H^\circ_{f, HCl \text{ in } 200H_2O} - \Delta H_{f, HCl(g)}$$

$$= -39.798 - (-22.06_2)$$

$$= -17.734 \text{ kcal mol}^{-1}$$

1.18 Calculate the enthalpies of reaction at 25 C for the following reactions in dilute aqueous solutions:

(a) $HC\ell\,(aq) + NaBr\,(aq) = HBr\,(aq) + NaC\ell\,(aq)$

(b) $CaC\ell_2(aq) + Na_2CO_3\,(aq) = CaCO_3(s) + 2\ NaC\ell\,(aq)$

SOLUTION:

(a) $\Delta H^\circ = 0$ because all of the reactants and products are completely ionized.

$$H^+ + C\ell^- + Na^+ + Br^- = H^+ + Br^- + Na^+ + C\ell^-$$

(b) $Ca^{2+}\,(aq) + CO_3^{\,2-}\,(aq) = CaCO_3\,(s)$

$$\Delta H^\circ = 288.45 - (-129.77) - (-161.63)$$

$$= 2.95 \text{ kcal mol}^{-1}$$

1.19 Calculate ΔH° for the dissociation

$$O_2\,(g) = 2O\,(g)$$

at 0, 298, and 3000 K. In Section 10.1 the

enthalpy change for dissociation at 0 K will be found to be equal to the spectroscopic dissociation energy D°.

SOLUTION:

ΔH° (0 K) = 2 (58.983) = 117.966 kcal mol^{-1}

ΔH° (298 K) = 2 (59.553) = 119.106 kcal mol^{-1}

ΔH° (3000 K) = 2 (61.358) = 122.716 kcal mol^{-1}

The spectroscopic dissociation energy of O_2 is given as 5.080 eV in Table 11.5. This can be converted to kcal mol^{-1} by multiplying by 23.06 kcal mol^{-1} eV^{-1}.

1.21 What is the heat evolved in freezing water at −10°C given that

$$H_2O(\ell) = H_2O(s) \quad \Delta H^\circ(273\,K) = -1435 \text{ cal mol}^{-1}$$

and $C_P(H_2O,\ell) = 18$ cal K^{-1} mol^{-1} and $C_P(H_2O,s) =$ 8.8 cal K^{-1} mol^{-1}

SOLUTION:

$$\Delta H^\circ(263\,K) = \Delta H^\circ(273\,K) + [C_{P,H_2O(s)} - C_{P,H_2O(\ell)}] \times (263\,K - 273\,K)$$

$$= -1435 \text{ cal mol}^{-1} + (-9.20 \text{ cal K}^{-1}\text{ mol}^{-1})(-10)$$

$$= -1343 \text{ cal mol}^{-1}$$

1.22 914 m

1.23 0.929 cal mol^{-1}

1.24 (a) 476 cal mol^{-1} (b) −5570 cal mol^{-1}

(c) −5570 cal mol^{-1} (d) −5094 cal mol^{-1}

1.25

	CO	CO_2	NH_3	CH_4
Cp (classical)	8.942	14.903	19.870	25.831
Cp (Table A.2 at 3000 K)	8.895	14.873	19.000	24.233

(The values of C_P are in cal K^{-1} mol^{-1})

1.26 9769 cal mol^{-1}

1.27 (a) 3.125 kcal mol^{-1} (b) 2.533 kcal mol^{-1}

1.28 (a) −73.5 L atm mol^{-1} (b) 1780 cal mol^{-1}

(c) 0 (d) 0

1.29 (a) 567 K (b) 9.30 atm (c) 1321 cal mol^{-1}

1.30 (a) 0.489 (b) 0.303 atm

1.31 (a) 2390 (b) 1460 cal

1.32 $\alpha = 1/T$ $\kappa = 1/P$

1.34 0.324 nm

1.35 (a) 19.82, (b) 1621, (c) 2739 atm

1.37 (a) −60.57, (b) −59.68 kcal mol^{-1}

1.38 −9.92 kcal mol^{-1}

1.39 2.87 kcal mol^{-1}

1.40

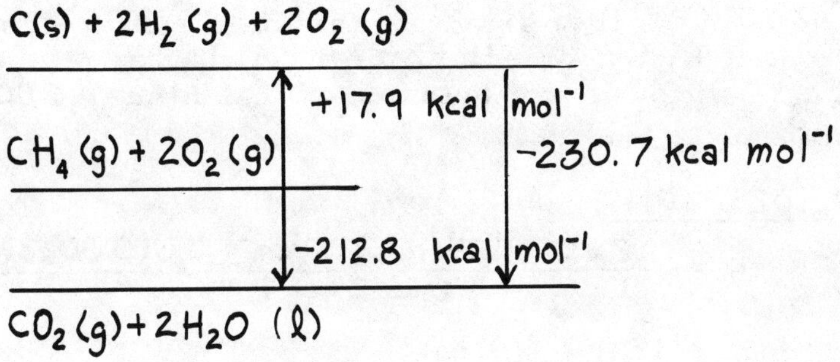

1.41 −139.9 ± 1.9 kcal mol^{-1} and −140.5 ± 0.4 kcal mol^{-1}

1.42 (a) −22.5, (b) −85.8 kcal mol^{-1}

1.43 (1) −28.9 (2) −12.0 (3) −4.76 (4) −10.8 kcal g^{-1}

1.44 −129.6, −44.124, −2.26 kcal mol^{-1}

1.45 53.957 kcal mol^{-1} is absorbed

1.46 25.521 kcal mol^{-1}

1.47 581.4 cal g^{-1}

CHAPTER 2. Second and Third Laws of Thermodynamics

2.1 Theoretically, how high could a gallon of gasoline lift an automobile weighing 2800 lb against the force of gravity, if it is assumed that the cylinder temperature is 2200 K and the exit temperature 1200 K? (Density of gasoline $= 0.80\ \mathrm{g\ cm^{-3}}$; $1\ \mathrm{lb} = 453.6\ \mathrm{g}$; $1\ \mathrm{ft} = 30.48\ \mathrm{cm}$; $1\ \mathrm{liter} = 0.2642\ \mathrm{gal}$. Heat of combustion of gasoline $= 11{,}200\ \mathrm{cal\,g^{-1}}$).

SOLUTION:

$$q = \frac{(11.2 \times 10^3\ \mathrm{cal\,g^{-1}})(1\ \mathrm{gal.})(10^3\ \mathrm{cm^{-3}\ L^{-1}})(0.80\ \mathrm{g\,cm^{-3}})}{0.2642\ \mathrm{gal.\ L^{-1}}}$$

$$= 3.39 \times 10^7\ \mathrm{cal}$$

$$w = q\,\frac{T_2 - T_1}{T_2} = \frac{(3.39 \times 10^7\ \mathrm{cal})(2200\ \mathrm{K} - 1200\ \mathrm{K})(4.184\ \mathrm{J\,cal^{-1}})}{(2200\ \mathrm{K})}$$

$$= 6.44 \times 10^{7}\ \mathrm{J}$$

$$= mgh = (2800\ \mathrm{lb})(0.4536\ \mathrm{kg\ lb^{-1}})(9.8\ \mathrm{m\,s^{-2}})$$

$$\times (0.3048\ \mathrm{m\,ft^{-1}})h$$

$$h = 17{,}000\ \mathrm{ft}$$

2.2 (a) What is the maximum work that can be obtained from 1000 cal of heat supplied to a water boiler at 100°C if the condenser is at 20°C?

(b) If the boiler temperature is raised to 150°C by the use of superheated steam under pressure, how much more work can be obtained?

SOLUTION:

(a) $w = q\frac{T_2 - T_1}{T_2} = (1000\ \text{cal})\frac{80\text{K}}{373.1\text{K}} = 214\ \text{cal}$

(b) $w = (1000\ \text{cal})\frac{130\text{K}}{423.1\text{K}} = 307\ \text{cal}$

or 93 cal more than (a)

2.3 Calculate the increase in entropy of a mole of silver that is heated at constant pressure from 0 to 30°C if the value of C_p in this temperature range is considered to be constant at 609 cal K^{-1} mol^{-1}.

SOLUTION:

$$\Delta S = C_p \ln\frac{T_2}{T_1} = (6.09\ \text{cal K}^{-1}\text{mol}^{-1})\ln\frac{303}{273}$$

$$= 0.635\ \text{cal K}^{-1}\text{mol}^{-1}$$

2.4 Calculate the change in entropy of a mole of aluminum which is heated from 600°C to 700°C. The melting point of aluminum is 660°C, the heat of fusion is 94 cal g^{-1}, and the heat capacities of the solid and liquid may be taken as 7.6 and 8.2 cal K^{-1} mol^{-1}, respectively.

SOLUTION:

$$\Delta S = \int_{T_1}^{T_f}\frac{C_{P,s}}{T}\,dT + \frac{\Delta H_f}{T_f} + \int_{T_f}^{T_2}\frac{C_{P,\ell}}{T}\,dT$$

$$= C_{P,s}\ln\frac{T_f}{T_1} + \frac{\Delta H_f}{T_f} + C_{P,\ell}\ln\frac{T_2}{T_f}$$

$$= (7.6\ \text{cal K}^{-1}\text{mol}^{-1})\ln\frac{933\text{K}}{873\text{K}} + \frac{(27\text{g mol}^{-1})(94\ \text{cal g}^{-1})}{933\text{K}}$$

$$+ (8.2\ \text{cal K}^{-1}\text{mol}^{-1})\ln\frac{973\text{K}}{933\text{K}}$$

$$= 355\ \text{cal K}^{-1}\text{mol}^{-1}$$

2.5 A mole of steam is condensed at 100°C and the water is cooled to 0°C and frozen to ice. What is the entropy change of the water? Consider that the average specific heat of liquid is 1.0 cal $K^{-1}g^{-1}$. The heat of vaporization at the boiling point and the heat of fusion at the freezing point are 539.7 and 79.7 calg^{-1}, respectively.

SOLUTION:

$$\Delta S = -\frac{\Delta H_{vap}}{T_{vap}} + \int_{373K}^{273K} \frac{C_p}{T}\, dT - \frac{\Delta H_{fus}}{T_{fus}}$$

$$= -\frac{(539.7\,\text{cal}\,g^{-1})(18.016\,g\,\text{mol}^{-1})}{373.15\,K}$$

$$+ (18.016\,\text{cal mol}^{-1}\,K^{-1}) \int_{373}^{273} d\ln T - \frac{(79.7\,\text{cal}\,g^{-1})(18.016\,g\,\text{mol}^{-1})}{273.15\,K}$$

$$= -36.9\ \text{cal}\ K^{-1}\ \text{mol}^{-1}$$

2.7 Calculate the entropy changes for the following processes: (a) melting of one mole of aluminum at its melting point, 660°C (ΔH_{fus} = 1.91 kcal mol^{-1}); (b) evaporation of mole of liquid oxygen at its boiling point, −182.97°C (ΔH_{vap} = 1.630 kcal mol^{-1}); (c) heating of one mol of hydrogen sulfide from 50 to 100°C at constant pressure (C_p = 7.15 + 0.00332T).

SOLUTION:

(a) $$\Delta S = \frac{\Delta H_{fus}}{T_{fus}} = \frac{1910\,\text{cal mol}^{-1}}{933K} = 2.05\ \text{cal}\ K^{-1}\ \text{mol}^{-1}$$

(b) $$\Delta S = \frac{\Delta H_{vap}}{T_{BP}} = \frac{1630\,\text{cal mol}^{-1}}{90.18K}$$

$$= 18.07\ \text{cal}\ K^{-1}$$

(c) $$\Delta S = \int_{T_1}^{T_2} \frac{C_p}{T}\, dT = \int_{323}^{373} \left(\frac{7.15}{T} + 0.00332\right) dT$$

$$= 7.15 \ln\frac{373}{323} + 0.00332\,(373-323)$$

$$= 1.195\ \text{cal}\ K^{-1}\ \text{mol}^{-1}$$

2.8 Calculate the entropy change in joules for a hundredfold isothermal expansion of a mole of perfect gas.

SOLUTION:

$$dS = \frac{dq_{rev}}{T} = \frac{P\,dV}{T} = \frac{R\,dV}{V}$$

$$\Delta S = R \ln \frac{V_2}{V_1} = (8.314 \text{ J K}^{-1} \text{ mol}^{-1}) \ln 100$$

$$= 38.3 \text{ J K}^{-1} \text{ mol}^{-1}$$

2.9 In the reversible isothermal expansion of a perfect gas at 300 K from 1 to 10 liters, where the gas has an initial pressure of 20 atm, calculate (a) ΔS for the gas and (b) ΔS for all systems involved in the expansion.

SOLUTION:

(a) $$n = \frac{PV}{RT} = \frac{(20 \text{ atm})(1 \text{ L})}{(0.08205 \text{ L atm K}^{-1} \text{ mol}^{-1})(300 \text{ K})}$$

$$= 0.812 \text{ mol}$$

$$\Delta S = nR \ln \frac{V_2}{V_1}$$

$$= (0.812 \text{ mol})(1.987 \text{ cal K}^{-1} \text{ mol}^{-1}) \ln \frac{10}{1}$$

$$= 3.72 \text{ cal K}^{-1}$$

(b) $\Delta S = 0$ since the process is carried out reversibly. The heat gained by the gas is equal to the heat lost by the heat reservoir, and both bodies are at the same temperature.

2.10 Calculate $\Delta G°$ for

$$H_2O(g, 25°C) = H_2O(l, 25°C)$$

The vapor pressure of water at 25°C is 23.76 torr.

SOLUTION:

$$\Delta G^\circ = -RT \ln \frac{a_{H_2O(\ell)}}{a_{H_2O(g)}}$$

$$= -(1.987 \text{ cal K}^{-1} \text{ mol}^{-1})(298.15 \text{ K}) \ln \frac{1}{(23.76/760)}$$

$$= -2.05 \text{ kcal mol}^{-1}$$

2.11 Calculate the change in Gibbs energy for the process

$$H_2O(\ell, -10^\circ C) = H_2O(s, -10^\circ C)$$

The vapor pressure of water at $-10^\circ C$ is 2.149 torr, and the vapor pressure of ice at $-10^\circ C$ is 1.950 torr. The process may be carried out by the following reversible steps:

1. A mole of water is transferred at $-10^\circ C$ from liquid to saturated vapor ($P = 2.149$ torr).

2. The water vapor is allowed to expand from 2.149 to 1.950 torr at $-10^\circ C$.

3. A mole of water is transferred at $-10^\circ C$ from vapor at $P = 1.950$ torr to ice at $-10^\circ C$.

SOLUTION:

Since the first and third steps are transfers at equilibrum, there are no Gibbs energy changes associated with them.

$$\Delta G_2 = -RT \ln \frac{P_1}{P_2}$$

$$= (1.987 \text{ cal K}^{-1} \text{ mol}^{-1})(263.15 \text{ K}) \ln \frac{2.149 \text{ torr}}{1.950 \text{ torr}}$$

$$= -50.8 \text{ cal mol}^{-1}$$

2.13 The standard entropy of H_2 (g) at 298 K is given in Table A.1. What is the entropy at 10 atm and 100 atm, assuming perfect gas behavior?

SOLUTION:

$$S = S^\circ - R \ln \frac{P}{P^\circ}$$

At 10 atm

$$S = 31.208 - (1.987) \ln 10$$

$$= 26.663 \text{ cal K}^{-1} \text{ mol}^{-1}$$

At 100 atm

$$S = 31.208 - (1.987) \ln 100$$

$$= 22.058 \text{ cal K}^{-1} \text{ mol}^{-1}$$

2.14 At 298 K, $S^\circ = 49.003$ cal K^{-1} mol^{-1} for O_2 (g). What is the entropy of O_2 (g) at 100 atm, assuming that it is a perfect gas?

SOLUTION:

$$S = S^\circ - R \ln \frac{P}{P^\circ} = 49.003 - 1.987 \ln 100$$

$$= 39.853 \text{ cal K}^{-1} \text{ mol}^{-1}$$

2.15 The change in Gibbs energy for the conversion of aragonite at 25°C is -250 cal mol^{-1}. The density of aragonite is 2.93 g cm^{-3} at 25°C and the density of calcite is 2.71 g cm^{-3}. At what pressure at 25°C would these two forms of $CaCO_3$ be in equilibrium?

SOLUTION:

aragonite = calcite $\Delta G^\circ = -250 \text{ cal K}^{-1} \text{ mol}^{-1}$

$$\Delta V = \frac{100.09 \text{ g mol}^{-1}}{2.71 \text{ g cm}^{-3}} - \frac{100.09 \text{ g mol}^{-1}}{2.73 \text{ g cm}^{-3}}$$

$$= 2.77 \text{ cm}^3 \text{ mol}^{-1} = 2.77 \times 10^{-3} \text{ L mol}^{-1}$$

$$\left(\frac{\partial \Delta G}{\partial P}\right)_T = \Delta V$$

$$\int_1^2 d\Delta G = \int_1^P \Delta V dP$$

$$\Delta G_2 - \Delta G_1 = \Delta V (P-1)$$

$$0 + 250 \text{ cal mol}^{-1} = (2.77 \times 10^{-3} \text{ L mol}^{-1})$$

$$\times \frac{(1.987 \text{ cal K}^{-1} \text{ mol}^{-1})}{(0.08205 \text{ L atm mol}^{-1})} (P-1)$$

$$P = 3780 \text{ atm}$$

2.17 From tables giving ΔG_f°, ΔH_f°, and C_P for H_2O (ℓ) and H_2O (v) at 298 K calculate (a) the vapor pressure of H_2O(ℓ) at 25°C and (b) the standard boiling point.

SOLUTION:

(a) $H_2O(\ell) = H_2O(g)$

$$\Delta G^\circ = -54.634 - (-56.687) = 2.053 \text{ kcal mol}^{-1}$$

$$\Delta G^\circ = -RT \ln (P/P^\circ)$$

$$P/P^\circ = \exp \frac{-2053 \text{ cal mol}^{-1}}{(1.987 \text{ cal K}^{-1} \text{ mol}^{-1})(298.15 \text{ K}}$$

$$= 3.126 \times 10^{-2}$$

$$P = (3.126 \times 10^{-2} \text{ atm})(760 \text{ torr atm}^{-1})$$

$$= 23.8 \text{ torr}$$

(b) $\Delta H°(298.15\,K) = -57.796 - (-68.315) = 10.519\ \text{kcal mol}^{-1}$

In the abence of data on the dependence of C_p on T we will calculate $\Delta H°(T)$ from

$$\Delta H°(T) = 10{,}519 + \int_{273.15}^{T} (8.025 - 17.995)\,dT$$

$$= 10{,}519 - 9.970\,(T - 273.15)$$

$$\frac{\partial(\Delta G°/T)}{\partial T} = -\frac{\Delta H°(T)}{T^2}$$

$$\frac{\Delta G°}{T} = -\int\left[\frac{-10{,}519}{T^2} - \frac{9.970}{T} + \frac{(9.970)(273.15)}{T^2}\right]dT$$

$$= \frac{10{,}519}{T} + 9.970\ln T + \frac{(9.970)(298.15)}{T} + I$$

where I is an integration constant which can be evaluated from $\Delta G°(298.15\,K)$.

$$\frac{2053}{298.15} = \frac{10519}{298.15} + 9.970\ln 298.15 + \frac{(9.970)(298.15}{(298.15} + I$$

$$I = -95.170$$

AT the boiling point $\Delta G° = 0$ and so we need to solve the following equation by successive approximations.

$$0 = \frac{10{,}519}{T} + 9.970\ln T + \frac{(9.970)(298.15)}{T} - 95.170$$

Trying T = 373 K

RHS = 0.039

Trying T = 374 K

RHS = −0.031

Therefore the standard boiling point calculated in this way is close to 373.5 K.

2.18 Calculate the entropy changes for the gas plus reservoir in Examples 2.5 and 2.6.

SOLUTION:

In Example 2.5 the gas plus the reservoir form an isolated system. Since the expansion is carried out reversibly, $\Delta S = 0$ for the isolated system.

In Example 2.6 no work is done and no heat flows to the reservoir. Therefore, the only entropy change in the isolated system is that of the gas, which is 4.58 cal K^{-1} mol^{-1}.

2.19 One mole of a perfect gas is allowed to expand reversibly and isothermally (25°) from a pressure of 1 atm to a pressure of 0.1 atm. (a) What is the change in Gibbs energy? (b) What would be the change in Gibbs energy if the process occurred irreversibly?

SOLUTION:

(a) Equation 2.71 $\left(\frac{\partial G}{\partial P}\right)_T = V = \frac{RT}{P}$

$$\Delta G = RT \ln \frac{P_2}{P_1}$$

$$= (1.987 \text{ cal K}^{-1} \text{ mol}^{-1})(298.15 \text{ K}) \ln 0.1$$

$$= -1364 \text{ cal mol}^{-1}$$

(b) $\Delta G = -1364$ cal mol^{-1} because G is a state function and depends only on the initial state and the final state.

2.21 On mole of a perfect gas at 300 K has an initial pressure of 15 atm and is allowed to expand isothermally to a pressure of 1 atm. Calculate (a) the maximum work that can be obtained from the expansion, (b) ΔU, (c) ΔH, (d) ΔG, (e) ΔS.

SOLUTION:

(a) $w = RT \ln \frac{P_1}{P_2} = (1.987 \text{ cal K}^{-1} \text{ mol}^{-1})(300 \text{ K}) \ln 15$

$= 1.614 \text{ kcal mol}^{-1}$

(b) $\Delta U = 0$ because $\partial U/\partial T = 0$ for a perfect gas

(c) $\Delta H = \Delta U + \Delta (PV) = 0$

(d) $\Delta G = RT \ln \frac{P_2}{P_1} = 1.614 \text{ kcal mol}^{-1}$

(e) $\Delta S = R \ln \frac{P_1}{P_2} = (1.987 \text{ cal K}^{-1} \text{mol}^{-1}) (\ln 15)$

$= 5.38 \text{ cal K}^{-1} \text{ mol}^{-1}$

2.22 One mole of amonia (considered to be a perfect gas) initially at 25°C+1 atm pressure is heated at constant pressure until the volume has trebled. Calculate (a) q, (b) w, (c) ΔH, (d) ΔV, (e) ΔS.

SOLUTION:

(a) $q = \int_{T_1}^{T_2} C_p \, dT$

$$= \int_{298}^{894} [6.189 + 7.887 \times 10^{-3} T - 7.28 \times 10^{-7} T^2] dT$$

$$= (6.189)(596) + \frac{1.887 \times 10^{-3}}{2} (894^2 - 298^2) - \frac{7.28 \times 10^{-3}}{3} (894^3 - 298^3)$$

$= 6320 \text{ cal mol}^{-1}$

(b) $w = -P\Delta V = -R(T_2 - T_1) = -(1.987 \text{ cal K}^{-1} \text{ mol}^{-1})(596 \text{ K})$

$= -1185 \text{ cal mol}^{-1}$

(c) $\Delta H = q_P = 6320 \text{ cal mol}^{-1}$

(d) $\Delta U = q + w = 6320 - 1185 = 5135$ cal mol^{-1}

(e) $$\Delta S = \int_{298}^{894} \frac{C_p\, dT}{T}$$

$$= \int_{298}^{894} \left[\frac{6.189}{T} + 7.887 \times 10^{-3} - 7.28 \times 10^{-7} T\right] dT$$

$$= 2.303\,(6.189) \log \frac{894}{298} + 7.887 \times 10^{-3}(894 - 298)$$
$$- \frac{7.28 \times 10^{-7}}{2}(894^2 - 298^2)$$

$$= 11.23 \text{ cal K}^{-1} \text{mol}^{-1}$$

2.23 Show that, by use of the second Law, the difference in heat capacities at constant pressure and constant volume

$$C_P - C_V = \left[P + \left(\frac{\partial U}{\partial V}\right)_T\right]\left(\frac{\partial V}{\partial T}\right)_P$$

may be written terms of quantities.

$$\alpha = \frac{1}{V}\left(\frac{\partial V}{\partial T}\right)_P \quad \text{and} \quad \kappa = -\frac{1}{V}\left(\frac{\partial V}{\partial P}\right)_T$$

that may be determined more easily from experiment, as

$$C_P - C_V = \frac{TV\alpha^2}{\kappa}$$

SOLUTION:

Since $dU = TdS - PdV$

$$\left(\frac{\partial U}{\partial V}\right)_T = T\left(\frac{\partial S}{\partial V}\right)_T - P$$

Using equation 2.83

$$\left(\frac{\partial U}{\partial V}\right)_T = T\left(\frac{\partial P}{\partial T}\right)_V - P$$

$$\therefore \quad C_P - C_V = T\left(\frac{\partial P}{\partial T}\right)_V \left(\frac{\partial V}{\partial T}\right)_P = T\,\frac{\left(\frac{\partial V}{\partial T}\right)_P^2}{\left(\frac{\partial V}{\partial P}\right)_T}$$

where the last form has been obtained using the cyclic rule from problem 2.41

$$\therefore \quad C_P - C_V = TV\alpha^2/\kappa$$

2.24 Calculate the partial molar volume of zinc chloride in 1- molar $ZnCl_2$ solution using the following data:

% by weight of $ZnCl_2$	2	6	10	14	18	20
Density $/\mathrm{g\,cm^{-3}}$	1.0167	1.0532	1.0891	1.1275	1.1665	1.1866

SOLUTION:

Taking the first solution as an an example, 1g of solution contains 0.02 g of $ZnCl_2$ ($M = 136.28\ \mathrm{g\,mol^{-1}}$) and 0.98 g of H_2O. The weight of solution containing 1000g of water is

$$\frac{1000}{0.98} = 1020\ \mathrm{g}$$

$$\text{molality } ZnCl_2 = \frac{(0.02)\,(1020\,\mathrm{g})}{136.28\,\mathrm{g\,mol^{-1}}}$$

$$= 0.1497\ \mathrm{m}$$

Volume of solution containing 1000g of H_2O

$$\frac{1020\,\mathrm{g}}{1.0167\,\mathrm{g\,cm^{-3}}} = 1003.2\ \mathrm{cm^3}$$

WT. %	Molality	Volume containing 1000g of H_2O
2	0.1497 m	1003.2
6	0.4683	1010.1

10	0.8152	1020.2
14	1.194	1031.3
18	1.610	1045.5
20	1.834	1053.4

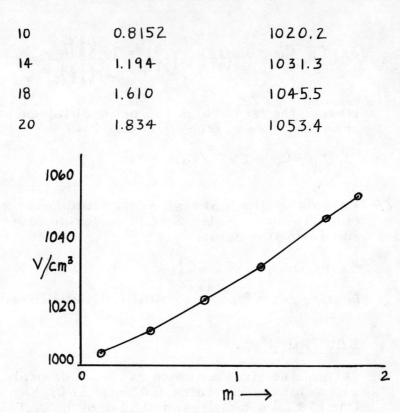

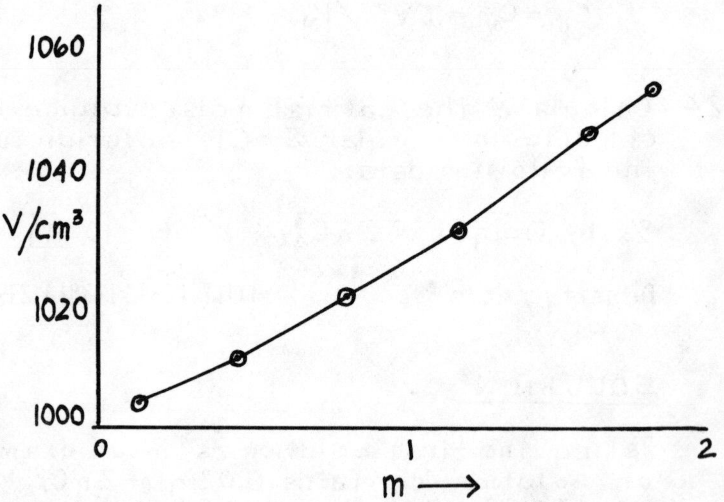

The slope of this plot at m=1 molar is 29.3 cm^3 mol^{-1} and so this is the partial molar volume of $ZnCl_2$.

2.26 Calculate the molar entropy of liquid chlorine at its melting point, 172.12 K, from the following data obtained by W. F. Giauque and T. M Powell:

T/K	15	20	25	30	35	40	50	60
C_p/cal $K^{-1}mol^{-1}$	0.89	1.85	2.89	3.99	4.97	5.73	6.99	8.00
T/K	70	90	110	130	150	170	172.12	
C_p/cal $K^{-1}mol^{-1}$	8.68	9.71	10.47	11.29	12.20	13.17	M.P	

The heat of fusion is 1531 cal mol^{-1}. Below 15 K it may be assumed that C_p is proportional to T^3.

SOLUTION:

For $T < 15$ K, $C_p = CT^3$, $C = \frac{0.89 \text{ cal K}^{-1}\text{mol}^{-1}}{(15 \text{ K})^3}$

$$S_{15K} = \int_0^{15 K} CT^3 \frac{dT}{T} = \left[CT^3/3\right]_0^{15}$$

$$= \frac{0.89}{15^3} \times \frac{15^3}{3} = 0.2967 \text{ cal K}^{-1}\text{ mol}^{-1}$$

The contributions from 15 K to the melting point are calculated using $C_p \Delta T/T$, where C_p is the average for the range and T is the average for the range.

K	Tavg	$C_{p,avg}$	ΔT
15-20	17.5	1.37	5
20-25	22.5	2.37	5
25-30	27.5	3.44	5
30-35	32.5	4.48	5
35-40	37.5	5.35	5
40-50	45	6.36	10
50-60	55	7.495	10
60-70	65	8.34	10
70-90	80	9.195	20
90-110	100	10.09	20
110-130	120	10.88	20
130-150	140	11.745	20
150-170	160	12.685	20
170-172.12	171.06	13.17	2.12

$$S_{liquid, T_m} = 0.2967 + \sum \frac{C_p \Delta T}{T} + \frac{\Delta H_{fus}}{T_m}$$

$$= 0.2967 + 16.56 + \frac{1531}{172.12}$$

$$= 25.75 \text{ cal K}^{-1} \text{ mol}^{-1}$$

2.27 Using molar entropies from Table A.1 calculate $\Delta S°$ for the following reactions at 25°C.

(a) $H_2(g) + \frac{1}{2} O_2(g) = H_2O(\ell)$

(b) $H_2(g) + C\ell_2(g) = 2HC\ell(g)$

(c) Methane (g) + $\frac{1}{2}$ O_2 (g) = methanol (ℓ)

SOLUTION:

(a) $\Delta S° = 16.71 - 31.208 - \frac{1}{2}(49.003)$

$= -39.00 \text{ cal K}^{-1} \text{ mol}^{-1}$

(b) $\Delta S° = 2(44.646) - 31.208 - 53.288$

$= 4.796 \text{ cal K}^{-1} \text{ mol}^{-1}$

(c) $\Delta S° = 30.3 - 44.492 - \frac{1}{2}(49.003)$

$= -38.7 \text{ cal K}^{-1} \text{ mol}^{-1}$

2.28 What is $\Delta S°$ for

$H_2(g) = 2H(g)$

at 298, 1000, and 3000 K?

SOLUTION:

$\Delta S°(298\text{ K}) = 2(27.391) - 31.208$

$= 23.574 \text{ cal K}^{-1} \text{ mol}^{-1}$

$\Delta S°(1000\text{ K}) = 2(33.403) - 39.702$

$= 27.104 \text{ cal K}^{-1} \text{ mol}^{-1}$

$$\Delta S^\circ (3000K) = 2(38.861) - 48.465$$
$$= 29.257 \text{ cal K}^{-1} \text{ mol}^{-1}$$

2.29 (a) 16.1% (b) 34.8% (c) 37.1%

(d) 37.3% (e) 47.1% (f) 62.0%

2.30 The gas is not returned to its initial state.

2.31 2.38 cal K^{-1} mol^{-1}

2.32 $\Delta S = R \ln [(V_2 - b)/(V_1 - b)]$

2.33 (a) 10.6 cal K^{-1} mol^{-1} (b) 7.8 cal K^{-1} mol^{-1}

2.34 29.97 cal K^{-1} mol^{-1}

2.35 $\Delta S_{system} = 0.005$ cal K^{-1} mol^{-1}

Therefore the change is spontaneous.

2.36 $\Delta S = -R(X_1 \ln X_1 + X_2 \ln X_2 + X_3 \ln X_3)$

$\Delta G = RT(X_1 \ln X_1 + X_2 \ln X_2 + X_3 \ln X_3)$

2.37 $dU = TdS - PdV = \left(\frac{\partial U}{\partial S}\right)_V dS + \left(\frac{\partial U}{\partial V}\right)_S dV$

$\therefore \left(\frac{\partial U}{\partial S}\right)_V = T$

$dH = TdS + VdP = \left(\frac{\partial H}{\partial S}\right)_P dS + \left(\frac{\partial H}{\partial P}\right)_S dP$

$$\therefore \left(\frac{\partial H}{\partial S}\right)_P = T$$

$$\left(\frac{\partial U}{\partial S}\right)_V = \left(\frac{\partial H}{\partial S}\right)_P$$

$$\therefore \left(\frac{\partial H}{\partial P}\right)_S = V$$

$$dG = -SdT + VdP = \left(\frac{\partial G}{\partial T}\right)_P dT + \left(\frac{\partial G}{\partial P}\right)_T dP$$

$$\therefore \left(\frac{\partial G}{\partial P}\right)_T = V$$

$$\left(\frac{\partial H}{\partial P}\right)_S = \left(\frac{\partial G}{\partial P}\right)_T$$

2.38 -27 cal mol^{-1}

2.39 21.4 cal mol^{-1}

2.40 $C_P - C_V = \dfrac{R^2 T V^3}{RTV^3 - 2a(V-b)^2}$

where V is the molar volume

2.42 $\left(\frac{\partial S}{\partial V}\right)_T = \left(\frac{\partial P}{\partial T}\right)_V = \frac{nR}{V-b}$

2.43 40.358 cal K^{-1} mol^{-1}

2.44 (a) 1194 cal mol^{-1} (b) 873 cal mol^{-1}

2.45 7730 cal mol^{-1}

2.46 (a) 1630 cal mol^{-1} (b) 1451 cal mol^{-1}

(c) 0 cal mol^{-1} (d) 18.1 cal K^{-1} mol^{-1}

2.47 (a) -1250 cal mol^{-1} (b) 1250 cal mol^{-1}

(c) 0 cal mol^{-1} (d) -1250 cal mol^{-1}

(e) 4.575 cal K^{-1} mol^{-1} (f) 0 cal mol^{-1}

(g) 0 cal mol^{-1} (h) 0 cal mol^{-1}

(i) -1250 cal mol^{-1} (j) 4.575 cal K^{-1} mol^{-1}

(k) 0 cal K^{-1} mol^{-1} (l) 4.575 cal K^{-1} mol^{-1}

2.48 741, -9720, -9720, -8979, 0, 741 cal mol^{-1}

-26.0 cal K^{-1} mol^{-1}

2.50 $\left(\frac{\partial U}{\partial V}\right)_T = \frac{a}{V^2}$

2.51 0.79 cm^3 g^{-1}

2.52 (a) 0.204 cm^3 g^{-1} (b) 19.4 cm^3 mol^{-1}

2.53 26.01 cm^3

2.54 -1131 cal mol^{-1} 3.79 cal K^{-1} mol^{-1}

2.55 27.6 cal K^{-1} mol^{-1}

2.56 36.04 cal K^{-1} mol^{-1}

2.57 69.0 cal K^{-1} mol^{-1}

2.59 (a) 19.235 cal K^{-1} mol^{-1} (b) -7.87 cal K^{-1} mol^{-1}

(c) -9.3 cal K^{-1} mol^{-1}

2.60 -19.28 cal K^{-1} mol^{-1}

The ions that are formed polarize neighboring water molecules and attract them in a hydration shell. This increases the order in the solution.

2.61 -1.21 cal K^{-1} mol^{-1} 10.857 cal K^{-1} mol^{-1}

CHAPTER 3. Phase Equilibria

3.1 What is the maximum number of phases that can be in equilibrium at constant temperature and pressure in one-, two-, and three-component systems?

SOLUTION:

$$v = c - p + 2$$

$$0 = 1 - p + 2, \quad p = 3$$

$$0 = 2 - p + 2, \quad p = 4$$

$$0 = 3 - p + 2, \quad p = 5$$

3.2 The critical temperature of carbon tetrachloride is 283.1°C. The densities in grams per cm^3 of the liquid ρ_l and vapor ρ_v at different temperatures are as follows:

t/°C	100	150	200	250	270	280
ρ_l	1.4343	1.3215	1.1888	0.9980	0.8666	0.7634
ρ_v	0.0103	0.0304	0.0742	0.1754	0.2710	0.3597

What is the critical molar volume of CCl_4? It is found that the mean of the densities of the liquid and vapor does not vary rapidly with temperature and can be represented by

$$\frac{\rho_l + \rho_v}{2} = AT + B$$

Where A and B are constants. The extrapolated value of the average density at the critical temperature is the critical density. The molar volume v_c at the critical point is equal to the molar mass divided by the critical density.

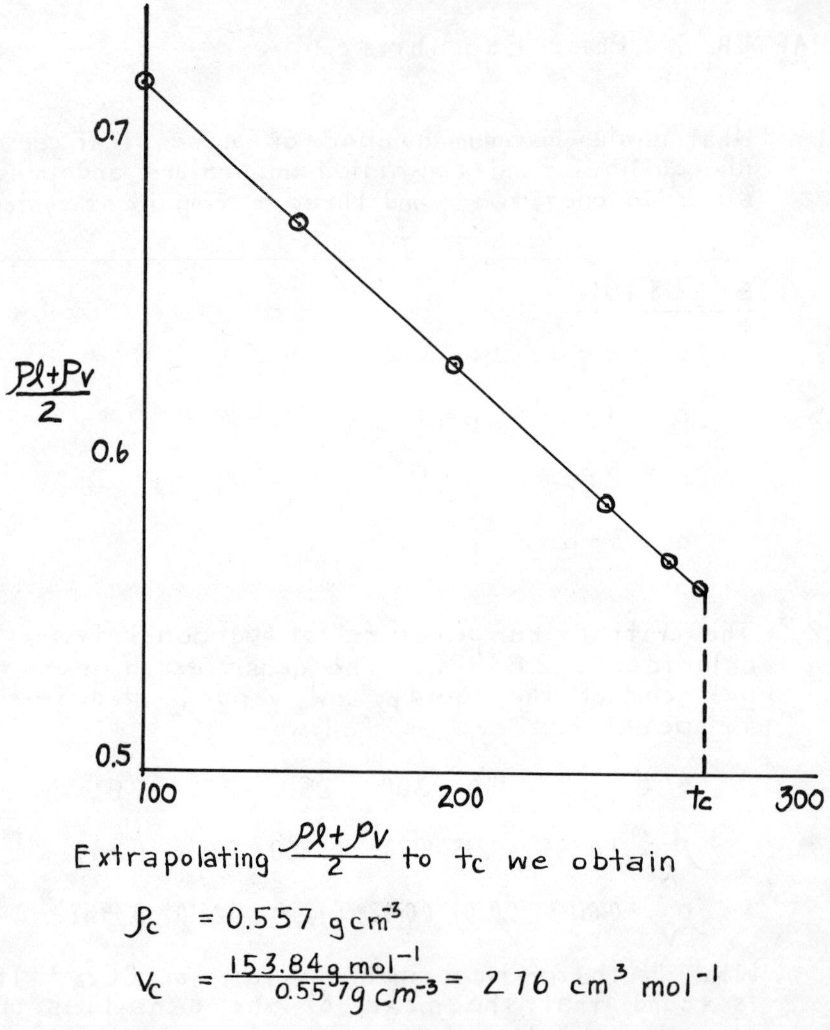

Extrapolating $\frac{\rho_l+\rho_v}{2}$ to t_c we obtain

$$\rho_c = 0.557 \text{ g cm}^{-3}$$

$$V_c = \frac{153.84 \text{ g mol}^{-1}}{0.557 \text{ g cm}^{-3}} = 276 \text{ cm}^3 \text{ mol}^{-1}$$

3.3 Liquid mercury has a density of 13.690 g cm^{-3}, and solid mercury has a density of 14.193 g cm^{-3}, both being measured at the melting point, $-38.87\,^\circ$C, under 1 atm pressure. The heat of fusion is 2.33 cal g^{-1}. Calculate the melting points of mercury under a pressure of (a) 10 atm and (b) 3540 atm. The observed melting point under 3540 atm is $-19.9\,^\circ$C.

SOLUTION:

$$\frac{\Delta T}{\Delta P} = \frac{T(V_l - V_s)}{\Delta H_{fus}} = \frac{(234.3\,K)\left(\frac{1}{13.690} - \frac{1}{14.193}\right) cm^3\, g^{-1}}{(2.33\, cal\, g^{-1})(41.29\, cm^3\, atm\, cal^{-1})}$$

$$= 0.00630\ K\ atm^{-1}$$

(a) $\Delta T = (0.00630\ K\ atm^{-1})(9\ atm) = 0.06°$

$t = -38.87° + 0.06° = -38.81°C$

(b) $\Delta T = (0.00630\ K\ atm^{-1})(3539\ atm) = 22.3°$

$t = -38.87° + 22.3° = 17°C$

3.4 The heats of vaporization and of fusion of water are 595 cal g^{-1} and 79.7 cal g^{-1} at 0°C. The vapor pressure of water at 0°C is 4.58 torr. Calculate the sublimation pressure of ice at −15°C, assuming that the enthalpy changes are independent of temperature.

SOLUTION:

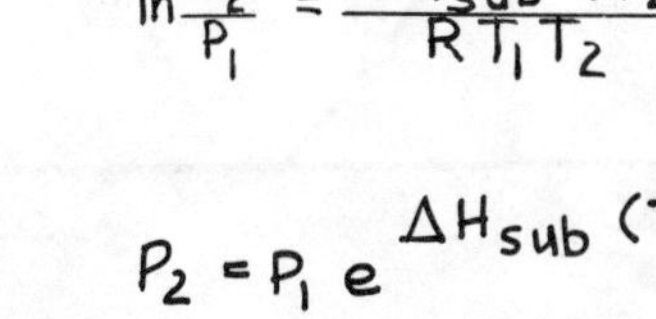

$$\Delta H_{sub} = \Delta H_{fus} + \Delta H_{vap}$$

$$= 79.7 + 595$$

$$= 675\ cal\ g^{-1}$$

$$\ln \frac{P_2}{P_1} = \frac{\Delta H_{sub}(T_2 - T_1)}{R T_1 T_2}$$

$$P_2 = P_1 e^{\Delta H_{sub}(T_2 - T_1)/RT_1T_2}$$

$$= (4.58\ torr)\, e^{\frac{(675 \times 18\ cal\, mol^{-1})(-15K)}{(1.987\ cal\, K^{-1} mol^{-1})(273.15K)(258.15K)}}$$

$$= 1.247\ torr$$

3.5 n-Propyl alcohol has the following vapor pressures:

$t/°C$	40	60	80	100
P/torr	50.2	147.0	376	842.5

Plot these data so as to obtain a nearly straight line, and calculate (a) the heat of vaporization and (b) the boiling point at 760 torr.

SOLUTION:

log p is plotted versus 1/T

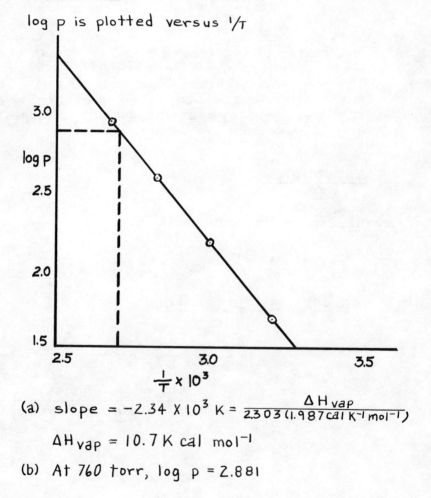

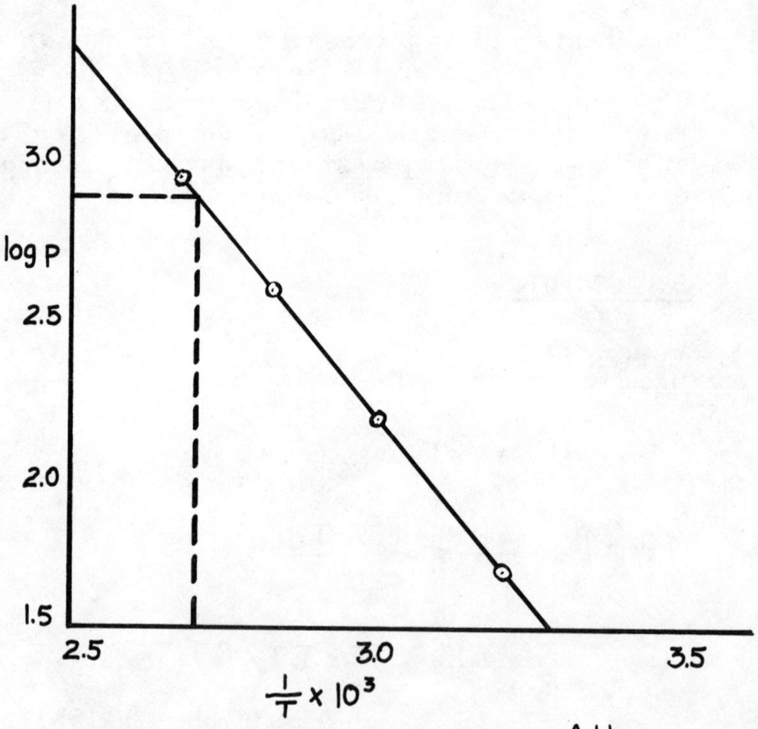

(a) $\text{slope} = -2.34 \times 10^3\ K = \dfrac{\Delta H_{vap}}{2.303(1.987\ cal\ K^{-1}\ mol^{-1})}$

$\Delta H_{vap} = 10.7\ K\ cal\ mol^{-1}$

(b) At 760 torr, $\log p = 2.881$

From the plot $\frac{1}{T} = 2.70 \times 10^{-3}$ $T = 370\,K$ or $97°C$

3.7 If $\Delta C_p = C_{p,vap} - C_{p,liq}$ is independent of temperature, then

$$\Delta H_{vap} = \Delta H_{0,vap} + T\Delta C_p$$

Where $\Delta H_{0,vap}$ is the hypothetical enthalpy of vaporization at absolute zero. Since ΔC_p is negative, ΔH_{vap} decreases as the temperature increases. Show that if the vapor is a perfect gas, the vapor pressure is given as a function of temperature by

$$\ln P = \frac{-\Delta H_{0,vap}}{RT} + \frac{\Delta C_p}{R} \ln T + \text{constant}$$

SOLUTION:

$$\frac{d \ln P}{dT} = \frac{\Delta H_{0,vap} + T\Delta C_p}{RT^2}$$

$$= \frac{\Delta H_{0,vap}}{RT^2} + \frac{\Delta C_p}{RT}$$

Integrating

$$\ln P = -\frac{\Delta H_{0,vap}}{RT} + \frac{\Delta C_p}{R} \ln T + \text{const}$$

3.8 The sublimation pressures of solid Cl_2 are 2.64 torr at $-112°C$ and 0.26 torr at $-126.5°C$. The vapor pressures of liquid Cl_2 are 11.9 torr at 100 and 58.7 torr at $-80°C$. Calculate (a) ΔH_{sub}, (b) ΔH_{vap}, (c) ΔH_{fus}, and (d) the triple point.

SOLUTION:

(a) $$\Delta H_{sub} = \frac{RT_1T_2}{T_2 - T_1} \ln \frac{P_2}{P_1}$$

$$= \frac{(1.987 \text{ cal K}^{-1}\text{ mol}^{-1})(161.15\text{ K})(146.65\text{ K})}{14.5\text{ K}} \ln \frac{2.64}{0.26}$$

$$= 7500 \text{ cal mol}^{-1}$$

(b) $$\Delta H_{vap} = \frac{(1.987 \text{cal K}^{-1}\text{ mol}^{-1})(173.15\text{ K})(193.15\text{K})}{20\text{ K}} \ln \frac{58.7}{11.9}$$

$$= 5300 \text{ cal mol}^{-1}$$

(c) $$\Delta H_{fus} = \Delta H_{sub} - \Delta H_{vap}$$

$$= 7500 - 5300 = 2200 \text{ cal mol}^{-1}$$

(d) For the solid

$$\ln P = \ln 2.64 + \frac{7500}{1.987}\left(\frac{1}{161.15} - \frac{1}{T}\right)$$

$$= 24.393 - \frac{3775}{T}$$

For the liquid

$$\ln P = \ln 11.9 + \frac{5300}{1.987}\left(\frac{1}{173.15} - \frac{1}{T}\right)$$

$$= 17.881 - \frac{2667}{T}$$

At the triple point

$$24.393 - \frac{3775}{T} = 17.881 - \frac{2667}{T}$$

$$T = \frac{1108}{6.512} = 170.2\text{ K}$$

$$T = 107.2\text{K} - 273.15\text{ K} = 103.0\,°\text{C}$$

3.10 Ethanol and methanol form very nearly ideal solutions. The vapor pressure of ethanol is 44.5 torr, and that of methanol is 88.7 torr, at 20° C. (a) Calculate the mole fraction of methanol and ethanol in a solution obtained by mixing 100 grams of each. (b) Calculate the partial pressures and the total vapor pressure of the solution. (c) Calculate the mole fraction of methanol in the vapor.

SOLUTION:

(a) $$X_{C_2H_5OH} = \frac{100/146}{100/146 + 100/32} = 0.410$$

$$X_{CH_3OH} = \frac{100/32}{100/146 + 100/32} = 0.590$$

(b) $$P_{C_2H_5OH} = X_{C_2H_5OH} \; P^\circ_{C_2H_5OH}$$

$$= (0.410)(44.5 \text{ torr}) = 18.2 \text{ torr}$$

$$P_{CH_3OH} = X_{CH_3OH} \; P^\circ_{CH_3OH}$$

$$= (0.590)(88.7 \text{ torr}) = 52.3 \text{ torr}$$

$$P_{total} = 18.2 \text{ torr} + 52.3 \text{ torr} = 70.5 \text{ torr}$$

(c) $$X_{CH_3OH,vapor} = \frac{52.3 \text{ torr}}{70.5 \text{ torr}} = 0.741$$

3.11 The vapor pressure of a solution containing 13 g of a nonvolatile solute in 100 g of water at 28° C is 27.371 torr. Calculate the molar mass of the solute, assuming that the solution is ideal. The vapor pressure of water at this temperature is 28.065 torr.

SOLUTION:

$$X_2 = \frac{n_2}{n_1+n_2} = \frac{P_1^\circ - P_1}{P_1^\circ} = \frac{28.065-27.371}{28.065} = 0.0247$$

$$\frac{n_2}{n_1+n_2} = 0.0247 = \frac{13/M}{(100/18)+(13/M)}$$

$$M = 92.4 \text{ g mol}^{-1}$$

3.12 Use the Gibbs-Duhem equation to show that if one component of a binary liquid solution follows Raoult's law, the other component will too.

SOLUTION:

$$X_1\, d\mu_1 + X_2\, d\mu_2 = 0 \qquad (1)$$

If $\mu_1 = \mu_1^\circ + RT \ln X_1$

$$d\mu_1 = \frac{RT}{X_1}\, dX_1$$

Using equation 1

$$d\mu_2 = -\frac{X_1}{X_2}\, d\mu_1 = \frac{RT}{X_2}\, dX_1$$

Since $X_1 + X_2 = 1$, $dX_2 = -dX_1$

$$d\mu_2 = \frac{RT}{X_2}\, dX_2 = RT\, d\ln X_2$$

$$\mu_2 = \text{const} + RT \ln X_2$$

If $X_2 = 1$, $\text{const} = \mu_2^\circ$

$$\mu_2 = \mu_2^\circ + RT \ln X_2$$

3.13 What are the entropy change and Gibbs energy change on mixing to produce a benzene-toluene solution with 1/3 mole fraction benzene at 25° C?

SOLUTION:

$$\Delta S_{mix} = -R\left(\tfrac{1}{3}\ln\tfrac{1}{3} + \tfrac{2}{3}\ln\tfrac{2}{3}\right)$$

$$= 1.265 \text{ cal K}^{-1}\text{ mol}^{-1}$$

$$\Delta G_{mix} = -T\Delta S_{mix}$$

$$= (298.15 \text{ K})(1.265 \text{ cal K}^{-1}\text{ mol}^{-1})$$

$$= -377 \text{ cal mol}^{-1}$$

3.15 The following table gives mole % acetic acid in aqueous solutions and in the equilibrium vapor at the boiling point of the solutions at 1 atm:

B.P./°C		118.1	113.8	107.5	104.4	102.1	100.0
Mole %	Liquid	100	90.0	70.0	50.0	30.0	0
acetic acid	Vapor	100	83.3	57.5	37.4	18.5	0

Calculate the minimum number of theoretical plates for the column required to produce an initial distillate of 28 mole % acetic acid from a solution of 80 mole % acetic acid.

SOLUTION:

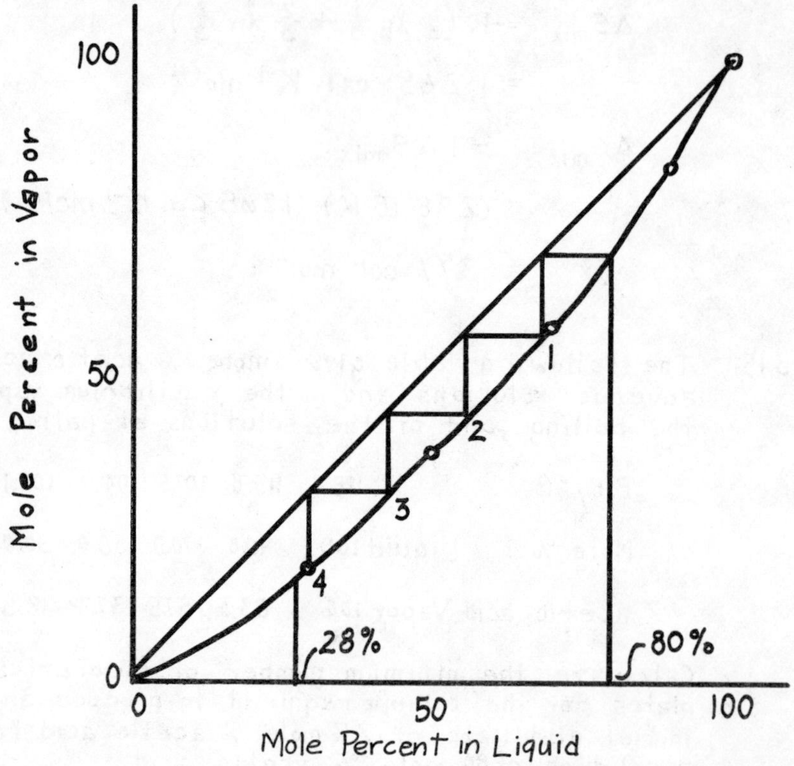

Since there are 4 steps, 3 theoretical plates are required in the column. The distilling pot counts as one plate.

3.16 If two liquids (1 and 2) are completely immiscible, the mixture will boil when the sum of the two partial pressures exceeds the applied pressure: $P = P_1° + P_2°$. In the vapor phase the ratio of the mole fractions of the two components is equal to the ratio of their vapor pressures.

$$\frac{P_1°}{P_2°} = \frac{X_1}{X_2} = \frac{g_1 M_2}{g_2 M_1}$$

Where g_1 and g_2 are the masses of components

1 and 2 in the vapor phase, and M_1 and M_2 are their molar masses. The boiling point of the immiscible liquid system napthalene-water is 98°C under a pressure of 733 torr. The vapor pressure of water at 98°C is 707 torr. Calculate the weight percent of napthalene in the distillate.

SOLUTION:

$$\frac{g_1}{g_2} = \frac{P_1^\circ M_1}{P_2^\circ M_2} = \frac{(733-707)(128)}{(707)(18)} = 0.261$$

$$\text{Weight percent napthalene} = \frac{0.261}{1.261} = 0.207 \text{ or } 20.7\%$$

3.17 Calculate the osmotic pressure of a 1 mol L^{-1} sucrose solution in water from the fact that at 30°C the vapor pressure of the solution is 31.207 torr. The vapor pressure of water at 30° C is 31.824 torr. The density of pure water at this temperature (0.99564 g cm^{-3}) may be used to estimate $\bar{V}_1$ for a dilute solution. To do this problem, Raoult's law is introduced into equation 3.56.

SOLUTION:

$$\bar{V}_1 \Pi = RT \ln X_1$$

Substituting Raoult's law $P_1 = X_1 P_1^\circ$

$$\Pi = -\frac{RT}{\bar{V}_1} \ln \frac{P_1}{P_1^\circ}$$

$$\bar{V}_1 = \frac{18.02 \text{ g mol}^{-1}}{0.99564 \text{ g cm}^{-3}} = 1810 \text{ cm}^3 \text{ mol}^{-1}$$

$$= 0.01810 \text{ L mol}^{-1}$$

$$\Pi = \frac{(0.08205 \text{ L atm K}^{-1}\text{mol}^{-1})(303.15 \text{ K})}{0.0180 \text{ L mol}^{-1}} \ln\frac{(31.824 \text{ torr})}{(31.207 \text{ torr})}$$

$$= 26.9 \text{ atm}$$

3.18 For a solution of n-propanol and water, the following partial pressures in torr are measured at 25°C. Draw a complete pressure-composition diagram, iincluding the total pressure. What is the composition of the vapor in equilibrium with a solution containing 0.5 mole fraction of n-propanol?

$X_{n\text{-propanol}}$	P_{H_2O}	$P_{n\text{-propanol}}$
0	23.76	0
0.020	23.5	5.05
0.050	23.2	10.8
0.100	22.7	13.2
0.200	21.8	13.6
0.400	21.7	14.2
0.600	19.9	15.5
0.800	13.4	17.8
0.900	8.13	19.4
0.950	4.20	20.8
1.000	0.00	21.76

SOLUTION:

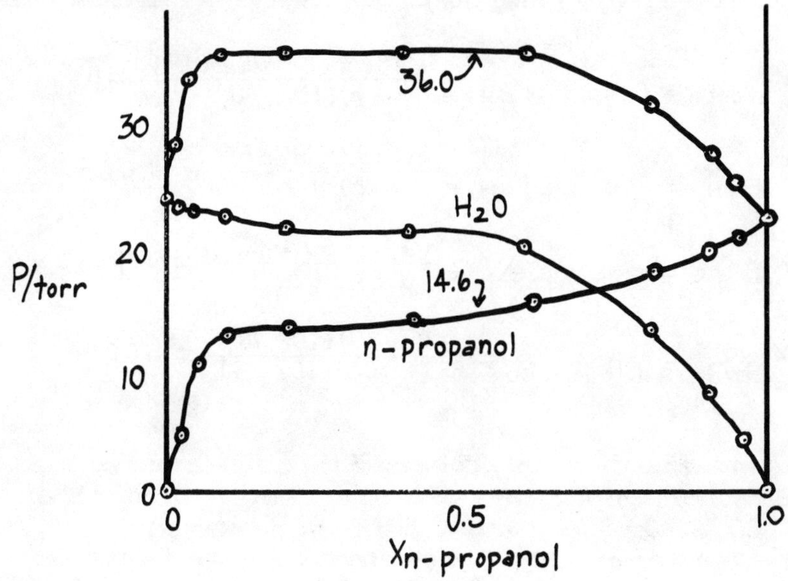

$$X_{vapor} = \frac{14.6 \text{ torr}}{36.0 \text{ torr}} = 0.406$$

3.20 The following data on ethanol-chloroform solutions at 35° C were obtained by G. Scatchard and C.L. Raymond [J. Am. Chem. Soc., 60, 1278 (1938)]:

X_{eto}, liq.	0	0.2	0.4	0.6	0.8	1.0
X_{eto}, vap.	0.0000	0.1382	0.1864	0.2554	0.4246	1.0000
Total pressure torr	295.11	304.22	290.20	257.17	190.19	102.78

SOLUTION:

$$X_{eto} = 0.2 \qquad f_{eto} = \frac{X_{vap}\, P_{vap}}{X_{liq.}\, P^{\circ}_{liq.}}$$

$$= \frac{(0.1382)(304.22)}{(0.2)(102.78)} = 2.045$$

$$X_{etoH} = 0.4 \quad f_{etoH} = \frac{(0.1864)(290.20)}{(0.4)(102.78)} = 1.316$$

$$X_{etoH} = 0.6 \quad f_{etoH} = \frac{(0.2554)(257.17)}{(0.6)(102.78)} = 1.065$$

$$X_{etoH} = 0.8 \quad f_{etoH} = \frac{(0.4246)(190.19)}{(0.8)(102.78)} = 0.982$$

$$X_{etoH} = 1.0 \quad f_{etoH} = \frac{(1.000)(102.78)}{(1.000)(102.78)} = 1.000$$

3.21 Using the data in problem 3.18, calculate the activity coefficients of water and n-propanol at 0.20, 0.40, 0.60, and 0.80 mole fraction n-propanol, using convention II and considering n-propanol to be the solvent.

SOLUTION:

Activity coefficients of n-propanol (Component 1)

$$X_1 = 0.2 \qquad \gamma_1 = \frac{13.6}{(0.2)(21.76)} = 3.13$$

$$X_1 = 0.4 \qquad \gamma_1 = \frac{14.2}{(0.4)(21.76)} = 1.63$$

$$X_1 = 0.6 \qquad \gamma_1 = \frac{15.5}{(0.6)(21.76)} = 1.19$$

$$X_1 = 0.8 \qquad \gamma_1 = \frac{17.8}{(0.8)(21.76)} = 1.02$$

To obtain the Henry law constant for H_2O (Component 2) plot P_2/X_2 versus X_2.

At $X_2 = 0.05, \frac{P_2}{X_2} = \frac{4.20}{0.05} = 84.0$

At $X_2 = 0.10, \frac{P_2}{X_2} = \frac{8.13}{0.1} = 81.3$

Thus as $X_2 \to 0$, $P_2 = 86.6 X_2$

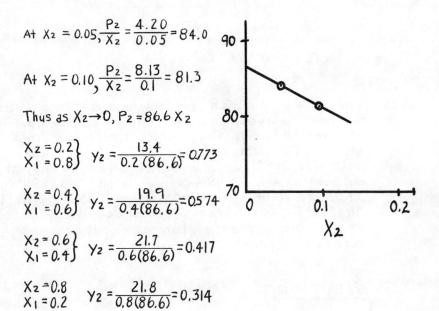

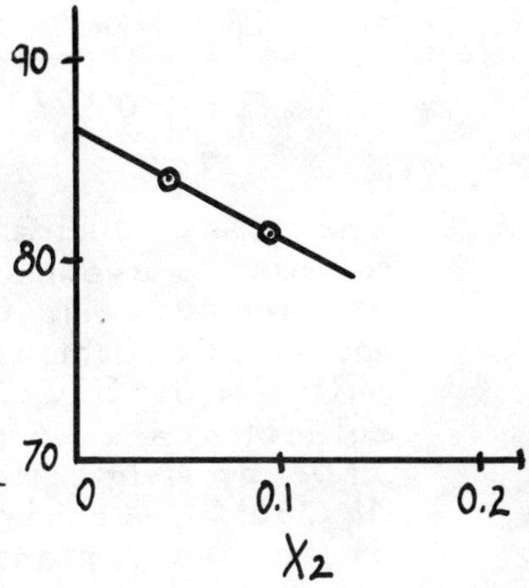

$\left.\begin{matrix} X_2 = 0.2 \\ X_1 = 0.8 \end{matrix}\right\} \quad \gamma_2 = \frac{13.4}{0.2\,(86.6)} = 0.773$

$\left.\begin{matrix} X_2 = 0.4 \\ X_1 = 0.6 \end{matrix}\right\} \quad \gamma_2 = \frac{19.9}{0.4(86.6)} = 0.574$

$\left.\begin{matrix} X_2 = 0.6 \\ X_1 = 0.4 \end{matrix}\right\} \quad \gamma_2 = \frac{21.7}{0.6(86.6)} = 0.417$

$\begin{matrix} X_2 = 0.8 \\ X_1 = 0.2 \end{matrix} \quad \gamma_2 = \frac{21.8}{0.8(86.6)} = 0.314$

3.23 If 68.4 g of sucrose ($M = 342$ g mol^{-1}) is dissolved in 1000 g of water: (a) What is the vapor pressure at 20° C? (b) What is the freezing point? The vapor pressure of water at 20° C is 17.363 torr.

SOLUTION:

(a) $$X_2 = \frac{\frac{68.4}{342}}{\frac{68.4}{342} + \frac{1000}{18}} = 3.59 \times 10^{-3}$$

$$\frac{P_1^\circ - P_1}{P_1^\circ} = X_2$$

$$\frac{17.363 - P_1}{17.363} = 3.59 \times 10^{-3}$$

$$P_1 = 17.363\,(1 - 3.59 \times 10^{-3})$$

$$= 17.300 \text{ torr}$$

(b) $\Delta T_f = K_f m = 1.86\ (0.2) = -0.372$

$T_f = -0.372°\ C$

3.24 The phase diagram for magnesium - copper at constant pressure shows that two compounds are formed: $MgCu_2$ which melts at 800°C, and Mg_2Cu, which melts at 580°C. Copper melts at 1085°C, and Mg at 648°C. The three eutectics are at 9.4% by weight Mg (680°C), 34% by weight Mg (560°C), and 65% by weight Mg (380°C). Construct the phase diagram. State the variance for each area and eutectic point.

SOLUTION:

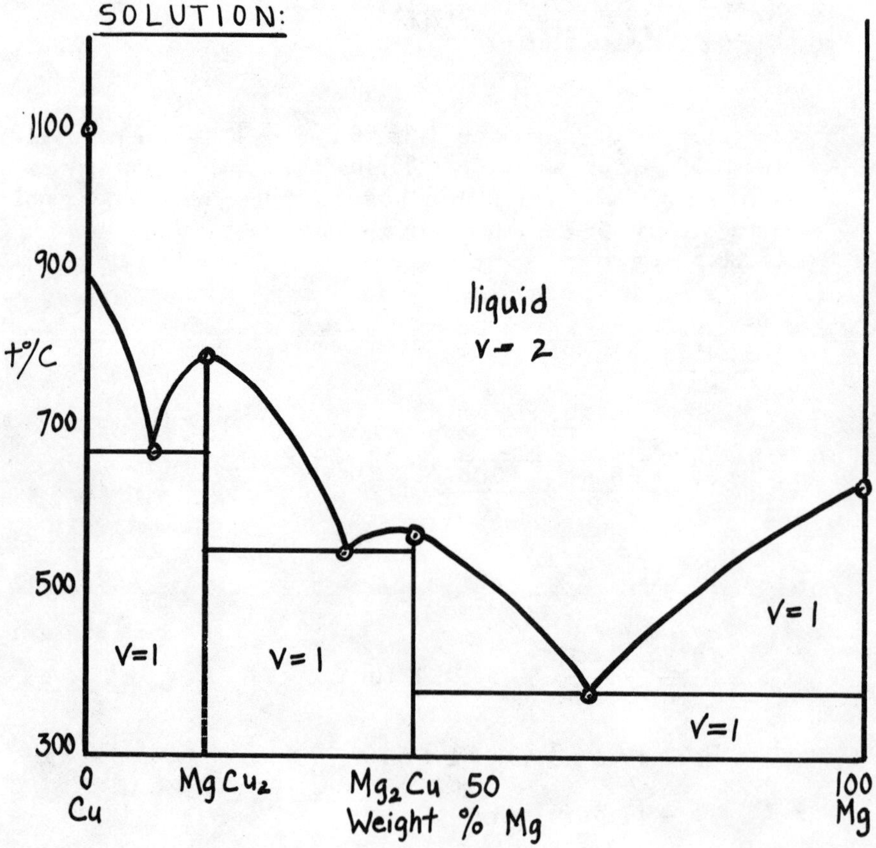

$v = 2 - p + 1$

In the liquid region $v = 2$, in the two-phase regions $v = 1$, and at the eutectic points $v = 0$

3.25 For the ternary system benzene-isobutanol at 25°C and 1 atm the following compositions have been obtained for the two phases in equilibrium:

Water-Rich Phase		Benzene-Rich Phase	
Isobutanol wt.%	Water wt.%	Isobutanol wt %	Benzene wt %
2.33	97.39	3.61	96.20
4.30	95.44	19.87	79.07
5.23	94.59	39.57	57.09
6.04	93.83	59.48	33.98
7.32	92.64	76.51	11.39

Plot these data on a triangular graph, indicating the tie lines. (a) Estimate the compositions of the phases that will be produced from a mixture of 20% isobutanol, 55% water, and 25% benzene. (b) What will be the composition of the principal phase when the first drop of the second phase separates when water is added to a solution of 80% isobutyl alcohol in benzene?

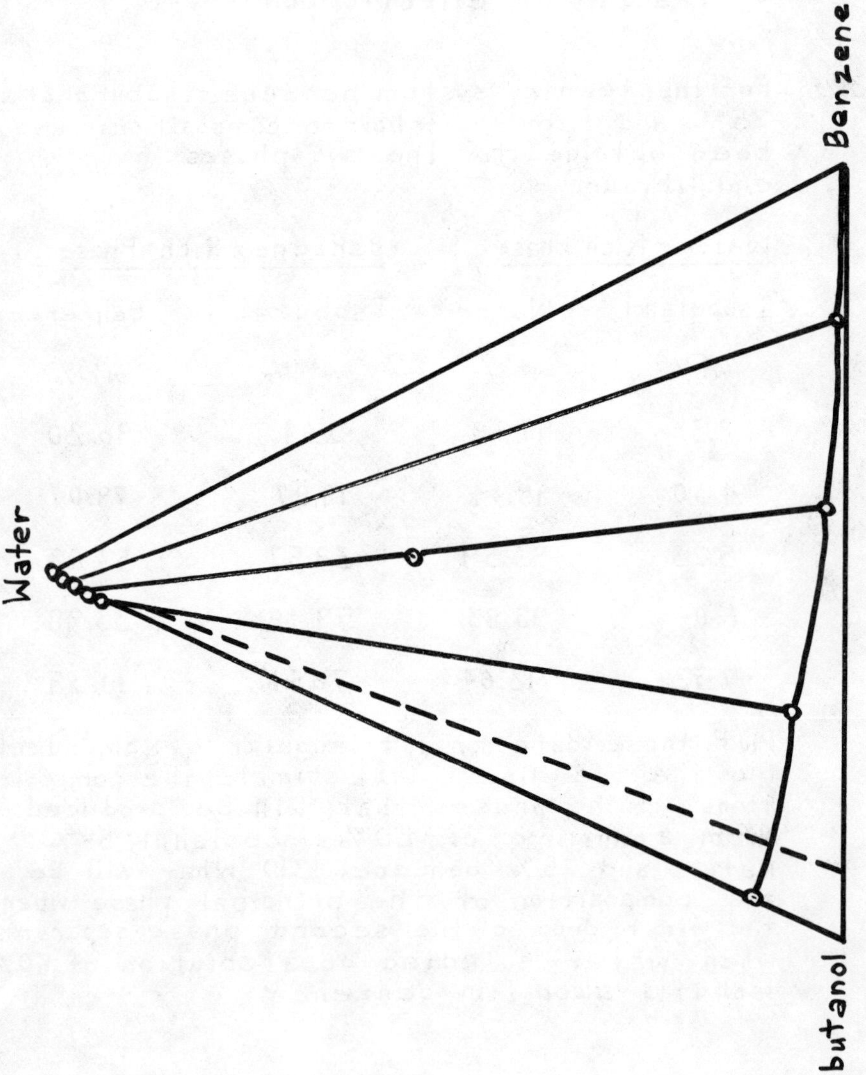

Water
Benzene
Isobutanol

3.26 The following data are available for the system nickel sulfate - sulfuric acid - water at 25° C. Sketch the phase diagram on triangular coordinate paper, and draw appropriate tie lines.

Liquid Phase		
$NiSO_4$, wt.%	H_2SO_4, wt.%	Solid Phase
28.13	0	$NiSO_4 \cdot 7H_2O$
27.34	1.79	$NiSO_4 \cdot 7H_2O$
27.16	3.86	$NiSO_4 \cdot 7H_2O$
26.15	4.92	$NiSO_4 \cdot 6H_2O$
15.64	19.34	$NiSO_4 \cdot 6H_2O$
10.56	44.68	$NiSO_4 \cdot 6H_2O$
9.65	48.46	$NiSO_4 \cdot H_2O$
2.67	63.73	$NiSO_4 \cdot H_2O$
0.12	91.38	$NiSO_4 \cdot H_2O$
0.11	93.74	$NiSO_4$
0.08	96.80	$NiSO_4$

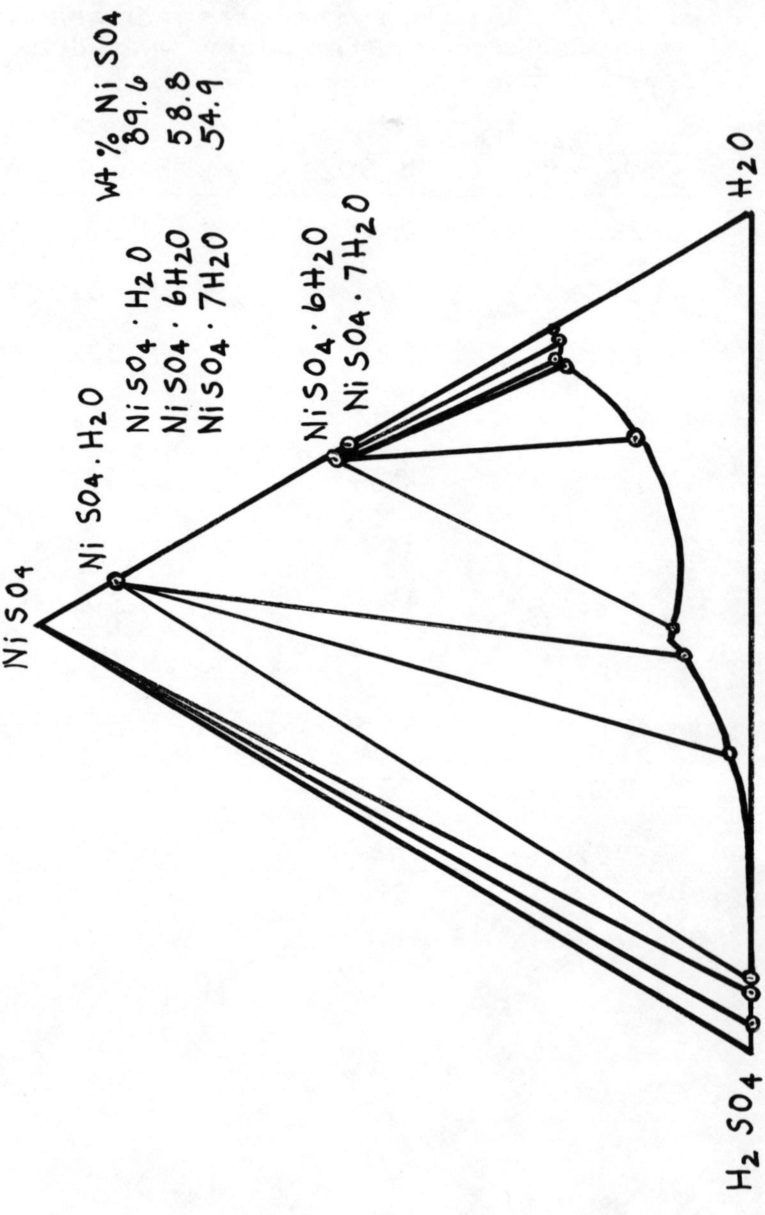

$NiSO_4$
$NiSO_4 \cdot H_2O$
$NiSO_4 \cdot 6H_2O$
$NiSO_4 \cdot 7H_2O$
H_2O
H_2SO_4
Wt % $NiSO_4$
$NiSO_4 \cdot H_2O$ 89.6
$NiSO_4 \cdot 6H_2O$ 58.8
$NiSO_4 \cdot 7H_2O$ 54.9

3.27 $0.274\ g\ cm^{-3}$ $131°\ C$

3.28 12.5 torr 1.25 torr

3.29 $96°C$

3.30 (a) $675\ cal\ g^{-1}$ (b) $0.376\ torr\ K^{-1}$

(c) 2.7 torr 2.9 torr

3.31 (a) 12,170 torr (b) 10,710 torr

3.32 $\Delta T = -5.4\ K$ Yes

3.33 5.7×10^{-4} torr

3.34 $6590\ cal\ mol^{-1}$

3.35 (a) $5.8°C$ 24.6 torr

(b) $2366\ cal\ mole^{-1}$

3.36 (a) $9110\ cal\ mol^{-1}$ (b) 28.4 torr

3.38 (a) $X_{CHCl_3} = 0.635$ $X_{CCl_4} = 0.365$

(b) 156.9 torr

3.39 (a) $X_{BrCH_2CH_2Br} = 0.802$

(b) $X_{BrCH_2CH_2Br} = 0.425$

3.40 (a) 23.713 torr (b) 23.713 torr

(c) 23.670 torr (d) 23.713 torr

3.41 (a) $X_B = 0.72$ $X_T = 0.28$ 521.5 torr

(b) $X_B = 0.537$

3.42 0.194 185.0 torr

3.43 (a) At −31.2 °C $X_{propane} = 0.560$

At −16.3 °C $X_{propane} = 0.196$

(b) At −31.2 °C $X_{propane, vap} = 0.885$

At −16.3 °C $X_{propane, vap} = 0.578$

3.44 (a) $X_{n\text{-propanol, vap}} = 0.37$

(b) $X_{n\text{-propanol, vap}} = 0.59$

3.45 $X_{C_6H_6} > 0.55$

3.46 562g

3.47 (a) $X_{EtOH} = 0.69$ (b) $X_{EtOH} = 1.00$

(c) $X_{EtOH} = 0.69$ (d) $X_{EtOH} = 0.88$

3.48 2.36×10^{-3} K

3.49

$X_{n\text{-propanol}}$	0.2	0.4	0.6	0.8
f_{H_2O}	1.14	1.52	2.09	2.82
$f_{n\text{-propanol}}$	0.248	0.129	0.094	0.081

3.50

X_{CHCl_3}	0.2	0.4	0.6	0.8
f_{CHCl_3}	0.58	0.70	0.84	0.96
$f_{acetone}$	0.98	0.89	0.74	0.61

3.51 $f_{acetone} = 1.67$ $\quad f_{CS_2} = 1.38$

3.52 0.108

3.53 $X_{Cd} = 0.844$ or 74.4 g/100 g of solution

3.54 251° C

3.55

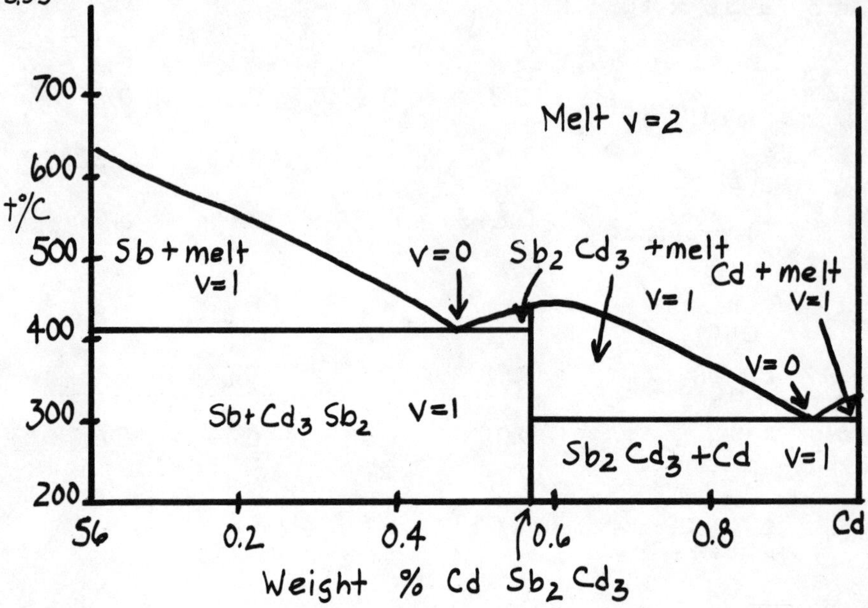

Melt v=2
Sb + melt
v=1
v=0
Sb2 Cd3 +melt
v=1
Cd + melt
v=1
v=0
Sb+Cd3 Sb2
v=1
Sb2 Cd3 +Cd
v=1
700
600
t°/C
500
400
300
200
Sb
0.2
0.4
0.6
0.8
Cd
Weight % Cd
Sb2 Cd3

3.56

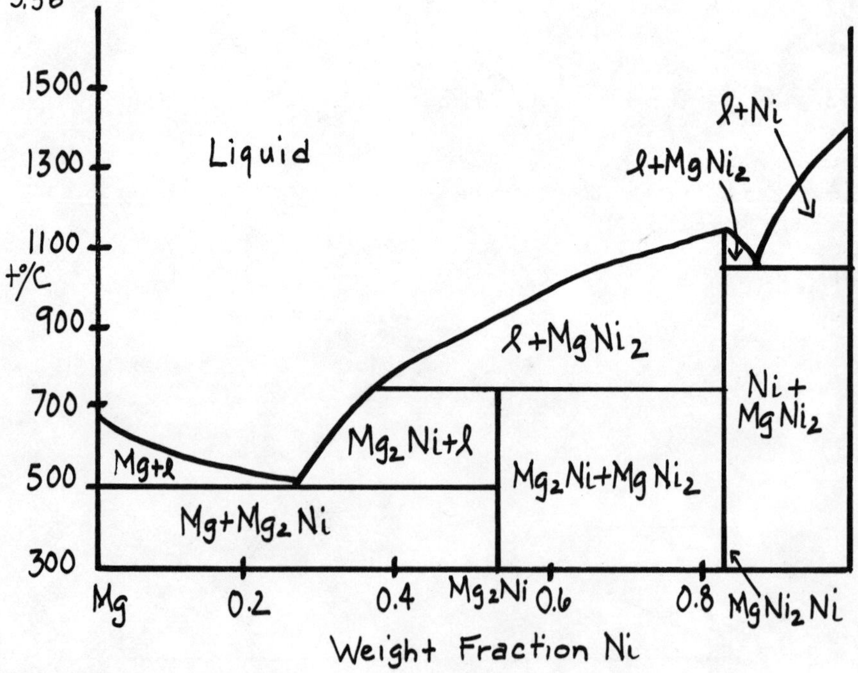

Liquid
l+Ni
l+MgNi2
l+MgNi2
Ni+
MgNi2
Mg2Ni+l
Mg+l
Mg2Ni+MgNi2
Mg+Mg2Ni
1500
1300
1100
t°/C
900
700
500
300
Mg
0.2
0.4
Mg2Ni
0.6
0.8
MgNi2
Ni
Weight Fraction Ni

3.57 See previous page

3.58

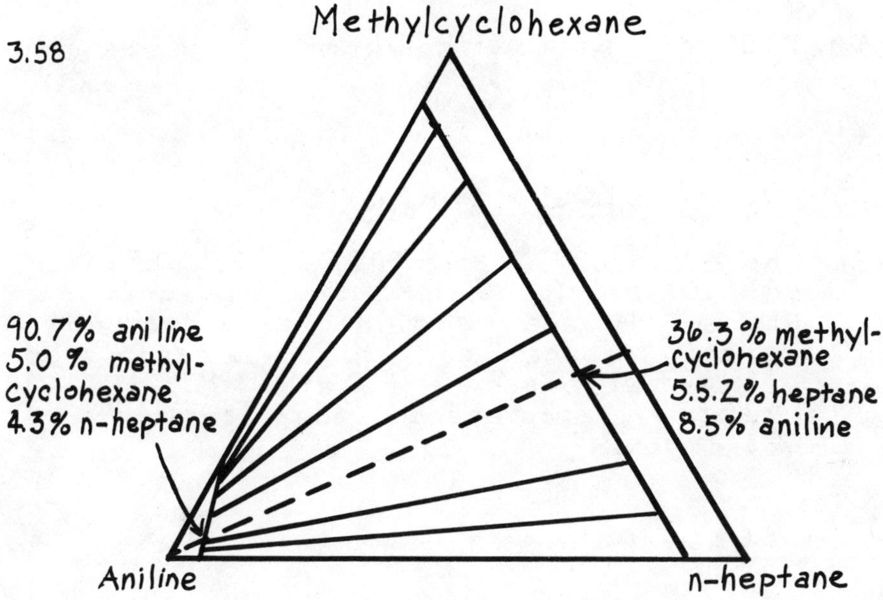

3.59

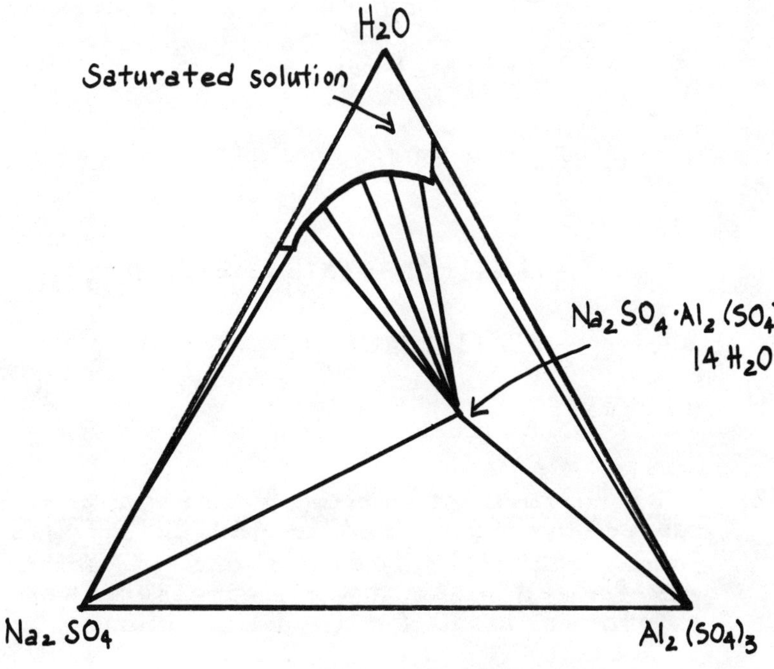

CHAPTER 4. Chemical Equilibrum

4.1 For the reaction

$$N_2 (g) + 3H_2 (g) = 2NH_3 (g)$$

$K_p = 1.64 \times 10^{-4}$ at 400 °C. Calculate (a) $\Delta G°$ and (b) ΔG when the pressures of N_2 and H_2 are maintained at 10 atm and 30 atm, respectively, and NH_3 is removed at a partial pressure of 3 atm. (c) Is the reaction spontaneous under the later conditions?

SOLUTION:

(a) $\Delta G° = -RT \ln K_p$

$$= -(1.987 \text{ cal K}^{-1} \text{mol}^{-1})(673 \text{ K}) \ln 1.64 \times 10^{-4}$$

$$= 11.66 \text{ kcal mol}^{-1}$$

(b) $$\Delta G° = \Delta G° + RT \ln \frac{P^2_{NH_3}}{P_{N_2} P^3_{H_2}}$$

$$= 11.66 + (1.987 \times 10^{-3})(673) \ln \frac{3^2}{10(30)^3}$$

$$= -2.13 \text{ kcal mol}^{-1}$$

(c) Yes

4.2 A 1:3 mixture of nitrogen and hydrogen was passed over a catalyst at 450 °C. It was found that 2.04 % by volume of ammonia was formed when the total pressure was maintained at 10 atm [A.T Larson and R.L. Dodge,

J. Am. Chem. Soc., 45, 2918 (1923)]. Calculate the value of K_p for $\frac{3}{2} H_2(g) + \frac{1}{2} N_2(g) = NH_3(g)$ at this temperature.

SOLUTION:

At equilibrium $P_{H_2} + P_{N_2} + P_{NH_3} = 10$ atm

$$P_{NH_3} = (10 \text{ atm})(0.0204) = 0.204 \text{ atm}$$

$$P_{H_2} + P_{N_2} = 10 \text{ atm} - 0.204 \text{ atm} = 9.896 \text{ atm}$$

$P_{H_2} = 3P_{N_2}$ because this is the initial ratio, and it is not changed by reaction.

$$P_{N_2} = \frac{9.896 \text{ atm}}{4} = 2.469 \text{ atm}$$

$$P_{H_2} = \frac{3}{4} 9.896 \text{ atm} = 7.407 \text{ atm}$$

$$K_p = \frac{(P_{NH_3}/P^\circ)}{(P_{H_2}/P^\circ)^{3/2} (P_{N_2}/P^\circ)^{1/2}}$$

$$= \frac{0.204}{7.407^{3/2} \; 2.469^{1/2}}$$

$$= 6.44 \times 10^{-3}$$

4.3 Water vapor is passed over coal (assumed to be pure graphite in this problem) at 1000K. Assuming that the only reaction occurring is the water gas reaction.

$$C(\text{graphite}) + H_2O(g) = CO(g) + H_2(g) \quad K_p = 2.49$$

calculate the equilibrium pressures of H_2O, CO, and H_2 at a total pressure of 1 atm. (Actually the water gas shift reaction

$$CO(g) + H_2O(g) = CO_2(g) + H_2(g)$$

occurs in addition, but it is considerably more complicated to take this subsequent reaction into account.)

SOLUTION:

$$K_p = \frac{(P_{CO}/P^\circ)(P_{H_2}/P^\circ)}{(P_{H_2O}/P^\circ)} = \frac{x^2}{y} = \frac{x^2}{1-2x} = 2.49$$

$$2x + y = 1$$

$$x^2 = 2.49 - 4.98x$$

$$x^2 + 4.98x - 2.49 = 0$$

$$x = \frac{-b \pm \sqrt{b^2 - 4ac}}{2a} = \frac{-4.98 + \sqrt{4.98^2 + 4(2.49)}}{2}$$

$$= 0.458 = (P_{CO}/P^\circ) = (P_{H_2}/P^\circ)$$

$$y = 1 - 2x = 0.084 = (P_{H_2O}/P^\circ)$$

$P_{H_2O} = 0.084$ atm $P_{CO} = 0.458$ atm $P_{H_2} = 0.458$ atm

4.4 How many moles of phosphorus pentachloride must be added to a liter vessel at 250°C to obtain a concentration of 0.1 mol of chlorine per liter? Given:

$$K_p = \frac{P_{PCl_3} P_{Cl_2}}{P_{PCl_5}} = 1.78$$

SOLUTION:

Let X represent the number of moles of PCl_5 added to the vessel

$$K_p = 1.78 = \frac{P_{PCl_3} P_{Cl_2}}{P_{PCl_5}} = \frac{(0.1RT/V)(0.1RT/V)}{(X-0.1)(RT/V)}$$

$$= \frac{0.1^2}{X-0.1}\left(\frac{RT}{V}\right)$$

$$1.78\,\text{atm} = \frac{(0.1\,\text{mol})^2}{X-0.1\,\text{mol}}\left[\frac{(0.08205\,\text{L atm K}^{-1}\,\text{mol}^{-1})(523.1\text{K})}{1\text{L}}\right]$$

$$X = 0.341\ \text{mol}$$

4.6 Rework Example 4.1 converting K_p to K_c using only SI units in the calculation.

SOLUTION:

$$K_c = K_p\left(\frac{P^\circ}{C^\circ RT}\right)^{\Sigma\nu_i}$$

$$= (1.64\times10^{-4})\left[\frac{101{,}325\ \text{N m}^{-2}}{10^3\ \text{mol m}^{-3})(8.314\,\text{J K}^{-1}\text{mol}^{-1})(673\text{K})}\right]^{-2}$$

$$= 0.500$$

4.7 At 1273 K and at a total pressure of 30 atm the equilibrium in the reaction $CO_2(g) + C(s) = 2\,CO(g)$ is such that 17 mol % of the gas is CO_2. (a) What percentage would be CO_2 if the total pressure where 20 atm? (b) What would be the effect on the equilibrium of adding N_2 to the reaction mixture in a closed vessel until the partial pressure of N_2 is 10 atm? (c) At what pressure of the reactantss will 25% of the gas be CO_2?

SOLUTION:

(a) $P_{CO_2} = (30\text{ atm})(0.17) = 5.1\text{ atm}$

$P_{CO} = (30\text{ atm})(0.83) = 24.9\text{ atm}$

$$K_P = \frac{(24.9)^2}{5.1} = 122$$

Let X = mole fraction CO_2

$$K_P = \frac{[20\ (1-X)]^2}{20X} = 122$$

$$X = 0.12$$

Percentage CO_2 at equilibrium = 12 %

(b) No effect for perfect gases because the partial pressures of the reactants are not affected.

(c)
$$K_P = \frac{[0.75\ (P/P^\circ)]^2}{0.25(P/P^\circ)} = 122$$

$$P = 54\text{ atm}$$

4.8 At 2000°C water is 2% dissociated into oxygen and hydrogen at a total pressure of 1 atm. (a) Calculate Kp. (b) Will the degree of dissociation increase or decrease if the pressure is reduced? (c) Will the degree of of dissociation increase or decrease if argon gas is added, holding the total pressure equal to 1 atm? (d) Will the degree of dissociation change if the pressure is raised by addition of argon at constant volume to the closed system containing partially dissociated water vapor? (e) Will the degree of dissociation increase or decrease if oxygen gas is added while holding the total pressure constant at 1 atm?

SOLUTION:

(a) $H_2O(g) = H_2(g) + \frac{1}{2} O_2(g)$

Initially	1	0	0	
Equilibrium	$1-\alpha$	α	$\alpha/2$	Total $= 1+\alpha/2$

$$P_{H_2O} = \frac{1-\alpha}{1+\frac{\alpha}{2}} P$$

$$P_{H_2} = \frac{\alpha}{1+\frac{\alpha}{2}} P$$

$$P_{O_2} = \frac{\alpha/2}{1+\frac{\alpha}{2}} P$$

$$K_P = \frac{\left[\frac{\alpha/2}{1+\alpha/2}\frac{P}{P^\circ}\right]^{1/2}\left[\frac{\alpha}{1+\alpha/2}\frac{P}{P^\circ}\right]}{\frac{1-\alpha}{1+\alpha/2}\frac{P}{P^\circ}} = \frac{\alpha^{3/2}(P/P^\circ)^{1/2}}{\sqrt{2}\,(1+\frac{\alpha}{2})^{1/2}(1-\alpha)}$$

$$= \frac{0.02^{3/2}\ 1^{1/2}}{\sqrt{2}\ (1.01)^{1/2}\ (0.98)} = 1.44$$

(b) If the total pressure is reduced, the degree of dissociation will increase because the reaction will produce more molecules to fill the volume.

(c) If argon is added at constant pressure, the degree of dissociation will increase because the partial pressure due to the reactants will decrease.

(d) If argon is added at constant volume the degree of dissociation will not be changed because the partial pressure due to the reactants will not change.

(e) If oxygen is added at total constant total pressure, the degree of dissociation of H_2O will decrease because the reaction will be pushed to the left.

4.9 In the synthesis of methanol by

$$CO(g) + 2H_2(g) = CH_3OH(g)$$

at 500 K, calculate the total pressure required for 90% conversion to methanol if CO and H_2 are initially in a 1:2 ratio. Given: $K_p = 6.25 \times 10^{-3}$.

SOLUTION:

$$CO(g) + 2H_2(g) = CH_3OH(g)$$

initial moles	1	2	0	
equil. moles	0.1	0.2	0.9	total 1.2

$$K_p = \frac{(P_{CH_3OH}/P^\circ)}{(P_{CO}/P^\circ)(P_{H_2}/P^\circ)^2} = 6.25 \times 10^{-3}$$

$$= \frac{\frac{0.9}{1.2}\frac{P}{P^\circ}}{\frac{0.1}{1.2}\frac{P}{P^\circ}\left(\frac{0.2}{1.2}\frac{P}{P^\circ}\right)^2}$$

$$\frac{P}{P^\circ} = \sqrt{\frac{(0.9)(1.2)^2}{(0.1)(0.04)(6.25 \times 10^{-3})}}$$

$$= 228$$

$P = 228$ atm = total pressure for 90% conversion to CH_3OH

4.10 An evacuated tube containing 5.96×10^{-3} mol L^{-1} of solid iodine is heated to 973K. The experimentally determined pressure is 0.490 atm [M.L. Perlman and G.K. Rollefson, J. Chem. Phys., 9, 362 (1941)] Assuming perfect gas behavior, calculate K_P for $I_2(g) = 2I(g)$.

SOLUTION:

$$P = \frac{n}{V}RT$$

$$0.490 \text{ atm} = (1+\alpha)(5.96 \times 10^{-3} \text{ mol L}^{-1})(0.08206 \text{ L atm K}^{-1} \text{ mol}^{-1})(973 \text{ K})$$

$$\alpha = \frac{0.490}{(5.96 \times 10^{-3})(0.08206)(973)} - 1$$

$$= 0.0297$$

$$K_P = \frac{4\alpha^2}{1-\alpha^2}\frac{P}{P^\circ} = \frac{4(0.0297)^2(0.490)}{1-0.0297^2}$$

$$= 1.73 \times 10^{-3}$$

4.12 For the reaction $N_2O_4(g) = 2NO_2(g)$, K_P at 25°C is 0.141. What pressure would be expected if 1 g of liquid N_2O_4 were allowed to evaporate into a liter vessel at this temperature? Assume that N_2O_4 and NO_2 are perfect gases.

SOLUTION:

$$K_P = \frac{4\alpha^2}{1-\alpha^2}\frac{P}{P^\circ} = 0.141$$

$$\frac{P}{P^\circ} = \frac{nRT}{VP^\circ}$$

$$n = \frac{(1 \text{ g L}^{-1})(1+\alpha)}{92 \text{ g mol}^{-1}}$$

$$K_p = \frac{4\alpha^2}{1-\alpha^2}\frac{(1 g L^{-1})(1+\alpha)(0.08206 \text{ L atm K}^{-1}\text{mol}^{-1})(298 \text{ K})}{(92 g \text{ mol}^{-1})(1 \text{ atm})}$$

$$\frac{\alpha^2}{1-d^2} = 0.1326$$

$$\alpha = 0.304$$

$$\frac{P}{P^o} = \frac{K_p(1-\alpha^2)}{4\alpha^2} = 0.346$$

$$P = 0.346 \text{ atm}$$

4.13 Under what total pressure at equilibrium must PCl_5 be placed at 250°C to obtain a 30% conversion into PCl_3 and Cl_2? For the reaction, $PCl_5(g) = PCl_3(g) + Cl_2(g)$, $K_p = 1.78$ at 250°C.

SOLUTION:

$$K_p = \frac{\alpha^2 (P/P^o)}{1 \alpha^2} = 1.78$$

$$\frac{P}{P^o} = \frac{K_p(1-\alpha^2)}{\alpha^2} = \frac{1.78(1-0.3^2)}{0.3^2}$$

$$P = 18.0 \text{ atm}$$

4.14 The equilibrium constant for the reaction $SO_2(g) + \frac{1}{2} O_2(g) = SO_3(g)$ at 727°C is given by

$$K_p = \frac{P_{SO_3}}{P_{SO_2} P_{O_2}^{1/2}} = 1.85$$

What is the ratio P_{SO_3}/P_{SO_2} (a) when the partial pressure of oxygen at equilibrium is 0.3 atm? (b) when the partial pressure of oxygen at equilibrium is 0.6 atm? (c) what is the effect on the equilibrium if the total pressure of the mixture of gases is increased by forcing in nitrogen at constant volume?

SOLUTION:

(a) $$\frac{P_{SO_3}}{P_{SO_2}} = K_P\left(\frac{P_{O_2}}{P^\circ}\right)^{1/2} = 1.85\,(0.3)^{1/2} = 1.01$$

(b) $$\frac{P_{SO_3}}{P_{SO_2}} = 1.85\,(0.6)^{1/2} = 1.43$$

(c) No effect if the gases behave as perfect gases.

4.15 A liter reaction vessel containing 0.233 mol of N_2 and 0.341 mol of PCl_5 is heated to 250°C. The total pressure at equilibrium is 28.95 atm. Assuming that all the gases are perfect, calculate K_P for the only reaction that occurs:

$$PCl_5\,(g) = PCl_3\,(g) + Cl_2\,(g)$$

SOLUTION:

$$\text{Total number of moles} = \frac{PV}{RT} = \frac{(28.95\text{ atm})(1\text{ L})}{(0.08206\text{ L atm K}^{-1}\text{mol}^{-1})(523.15\text{ K})}$$

$$= 0.674$$

Moles of reactants $= 0.674 - 0.233 = 0.441$

$= (0.341 - x) + 2x$

x = moles of PCl_3 = moles of Cl_2 = 0.100

$$K_P = \frac{(P_{PCl_3}/P^\circ)(P_{Cl_2}/P^\circ)}{P_{PCl_5}/P^\circ}$$

$$= \frac{\left(\frac{0.100}{0.674}\right)^2 (28.95)^2}{\left(\frac{0.241}{0.674}\right) 28.95}$$

$$= \frac{0.100^2 \; (28.95)}{(0.674)\;(0.241)}$$

$$= 1.78$$

4.17 Calculate (a) K_P and (b) $\Delta G°$ for the following reaction at 20°C:

$$CuSO_4 \cdot 4NH_3 (s) = CuSO_4 \cdot 2NH_3 (s) + 2NH_3 (g)$$

The equilibrium pressure of NH_3 is 62 torr.

(a) $K_P = (P_{NH_3}/P°)^2$

$$= \left(\frac{62 \text{ torr}}{-760 \text{ torr}}\right)^2 = 6.66 \times 10^{-3}$$

(b) $\Delta G° = -RT \ln K_P$

$$= -(1.987 \text{ cal K}^{-1} \text{ mol}^{-1})(293.15 \text{ K}) \ln 6.66 \times 10^{-3}$$

$$= 2920 \text{ cal mol}^{-1}$$

4.18 Air (20% O_2) will not oxidize silver at 200°C and 1atm pressure. Make a statement about the equilibrium constant for the reaction

$$2Ag(s) + \tfrac{1}{2}O_2(g) = Ag_2O(s)$$

SOLUTION:

$$K_P = \left(\frac{P_{O2}}{P°}\right)^{-1/2} = (0.5)^{-1/2} = 2.24 \text{ if}$$

Ag, O_2, and Ag_2O are in equilibrium. If they are not in equilibrium, a higher pressure of O_2 will be required and the value of K_p will be smaller than the above value: $K_p < 2.24$.

4.19 The dissociation of ammonium carbamate takes place according to the reaction

$$(NH_2)CO(ONH_4)(s) = 2NH_3(g) + CO_2(g)$$

When an excess of ammonium carbamate is placed in a previously evacuated vessel, the partial pressure generated by NH_3 is twice the partial pressure of the CO_2, and the partial pressure of $(NH_2)CO(ONH_4)$ is negligible in comparison. Show that

$$K_p = \frac{4}{27}(P/P^\circ)^3$$

where P is the total pressure.

SOLUTION:

$$P = P_{NH_3} + P_{CO_2} = 3P_{CO_2} \qquad \text{since } P_{NH_3} = 2P_{CO_2}$$

$$P_{CO_2} = P/3 \qquad P_{NH_3} = \frac{2}{3}P$$

$$K_p = \left(\frac{P_{NH_3}}{P^\circ}\right)^2\left(\frac{P_{CO_2}}{P^\circ}\right) = \left(\frac{2}{3}\frac{P}{P^\circ}\right)^2\left(\frac{1}{3}\frac{P}{P^\circ}\right)$$

$$= \frac{4}{27}\left(\frac{P}{P^\circ}\right)^3$$

4.20 From the $\Delta G°_f$ of Br_2 (g) at 25°C, calculate the vapor pressure of Br_2 (l). The pure liquid at 1atm and 25° C is taken as the standard state.

SOLUTION:

$$Br_2(l) = Br_2(g) \qquad K_p = P_{Br_2}/P°$$

$$\Delta G° = -RT \ln (P_{Br_2}/P°)$$

$$\frac{P_{Br_2}}{P°} = e^{\frac{-751 \text{ cal mol}^{-1}}{(1.987 \text{ cal K}^{-1} \text{mol}^{-1})(298.15 \text{ K})}}$$

$$= 0.281$$

$$P_{Br_2} = (0.281 \text{ atm})(760 \text{ torr atm}^{-1})$$

$$= 214 \text{ torr}$$

4.22 Calculate the equilibrium constants for the following reactions at 25°.

(a) $2CH_4(g) = C_2H_6(g) + H_2(g)$

(b) $CH_4(g) + H_2O(g) = CH_3OH(g) + H_2(g)$

(c) $C_2H_6(g) = C_2H_4(g) + H_2(g)$

(d) $C_2H_4(g) + H_2O(g) = C_2H_5OH(g)$

SOLUTION:

(a) $\Delta G° = -7.86 - 2(-12.13) = 16.40 \text{ kcal mol}^{-1} = -RT \ln K$

$$K = e^{-16,400/(1.987)(298)} = 9.36 \times 10^{-13}$$

(b) $\Delta G° = -38.72 - (-12.13) - 54.63 = -81.22 \text{ kcal mol}^{-1}$

$K = 3.72 \times 10^{59}$

(c) $\Delta G° = 16.28 - (-7.86) = 24.14 \text{ kcal mol}^{-1}$

$K = 1.97 \times 10^{-18}$

(d) $\Delta G° = -40.29 - 16.28 - (-54.63) = -1.94 \text{ kcal mol}^{-1}$

$K = 26.5$

4.23 In order to produce more hydrogen from "synthesis gas" ($CO + H_2$) the water gas shift reaction is used.

$$CO(g) + H_2O(g) = CO_2(g) + H_2(g)$$

Calculate K_P at 1000 K and the equilibrium extent of reaction starting with an equimolar mixture of CO and H_2O.

SOLUTION:

$$\Delta G° = -94,628 + 47,859 + 46,040 = -729 \text{ cal mol}^{-1}$$

$$= -(1.987 \text{ cal K}^{-1} \text{mol}^{-1})(1000 \text{K}) \ln K_P$$

$$K_P = 1.44 = \frac{P_{H_2} P_{CO_2}}{P_{CO} P_{H_2O}} = \frac{x^2}{(1-x)^2}$$

$x = 0.545$, fractional conversion of reactants to products

(Note that this reaction is exothermic so that there will be a larger extent of reaction at lower temperatures. In practice this reaction is usually carried out at about 700 K.)

4.24 Calculate the equilibrium constant Kp for the production of H_2 from CH_4 at 1000 K using the reaction.

$$CH_4(g) + H_2O(g) = CO(g) + 3H_2(g)$$

SOLUTION:

$$\Delta G° = -47{,}859 - 4{,}625 + 46{,}040 = -6444 \text{ cal mol}^{-1}$$

$$= -(1.987 \text{ cal K}^{-1}\text{ mol}^{-1})(1000\text{ K}) \ln K_p$$

$$K_p = 2.56$$

4.25 Calculate the equilibrium constants for

$$C(\text{graphite}) + 2H_2(g) = CH_4(g)$$

at 500 K and 1000 K.

SOLUTION:

$$\Delta G°_{500\text{ K}} = -7845 \text{ cal mol}^{-1}$$

$$= -(1.987 \text{ cal K}^{-1}\text{ mol}^{-1})(500\text{ K}) \ln K_p$$

$$K_p = 2.69 \times 10^3$$

$$\Delta G°_{1000\text{ K}} = 4625 \text{ cal mol}^{-1}$$

$$= -(1.987 \text{ cal K}^{-1}\text{ mol}^{-1})(1000\text{ K}) \ln K_p$$

$$K_p = 9.75 \times 10^{-2}$$

2.46 What are the percentage dissociations of $H_2(g)$, $O_2(g)$, and $I_2(g)$ at 2000 K and a total pressure of 1 atm?

SOLUTION:

$H_2(g) = 2H(g)$

$$K_p = \frac{(P_H/P^\circ)^2}{P_{H_2}/P^\circ} = \frac{4\alpha^2 (P/P^\circ)}{1-\alpha^2}$$

$$\Delta G^\circ = -RT \ln K_p$$

$$= 2(25{,}543) = -(1.987)(2000) \ln K_p$$

$$K_p = 2.61 \times 10^{-6} = \frac{4\alpha^2}{1-\alpha^2}$$

$\alpha = 8.08 \times 10^{-4}$ or 0.081 %

$O_2(g) = 2O(g)$

$$\Delta G^\circ = 2(29{,}079) = -(1.987)(2000) \ln K_p$$

$$K_p = 4.41 \times 10^{-7} = \frac{4\alpha^2}{1-\alpha^2}$$

$\alpha = 3.32 \times 10^{-4}$ or 0.033 %

$I_2(g) = 2I(g)$

$$\Delta G^\circ = 2(-7031) = -(1.987)(2000) \ln K_p$$

$$K_p = 34.42 = \frac{4\alpha^2}{1-\alpha^2}$$

$\alpha = 0.947$ or 94.7%

4.28 The following data apply to the reaction $Br_2(g) = 2Br(g)$:

T/K	1123	1173	1223	1273
$K_p/10^{-3}$	0.403	1.40	3.28	7.1

Determine by graphical means the enthalpy change when 1 mol of Br_2 dissociates completely at 1200 K.

SOLUTION:

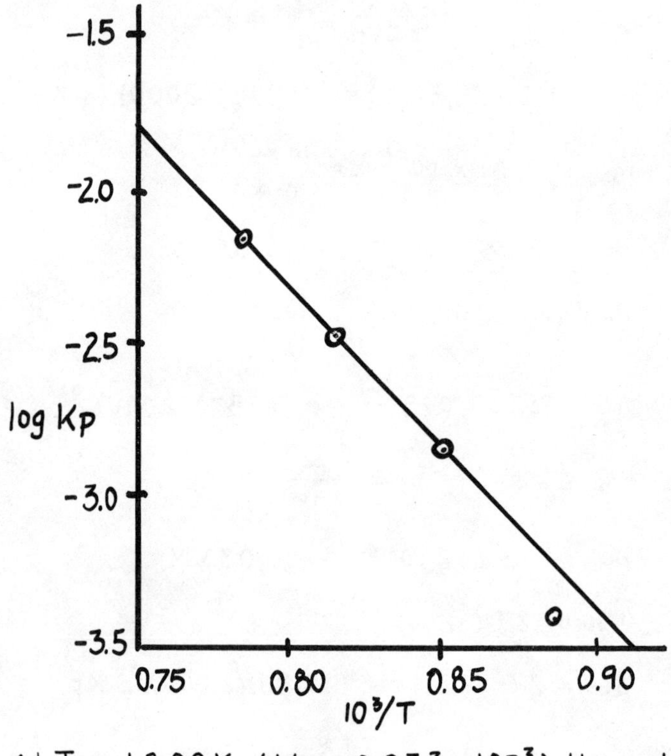

At T = 1200 K ($1/T = 0.833 \times 10^{-3}$) the slope is -10.4×10^3 K

$$\Delta H^\circ = -2.303\, R\, (\text{slope})$$

$$= (2.303)(1.987 \text{ cal K}^{-1} \text{ mol}^{-1})(10.4 \times 10^3 \text{ K})$$

$$= 47.6 \text{ Kcal mol}^{-1}$$

4.29 Mercuric oxide dissociates according to the reaction $2HgO(s) = 2Hg(g) + O_2(g)$. At 420 °C the dissociation pressure is 387 torr, and at 450 °C it is 810 torr. Calculate (a) the equilibrium constants, and (b) the enthalpy of dissociation per mole of HgO.

SOLUTION:

(a) $P_{Hg} = 2P_{O_2}$ $\quad P_{Hg} = \frac{2}{3}P$ $\quad P_{O_2} = \frac{1}{3}P$

$$K_{P420} = P_{Hg}P_{O_2} = \left(\frac{2}{3}\right)^2\left(\frac{1}{3}\right)P^3$$

$$= \left(\frac{4}{27}\right)\left(\frac{387\text{ torr}}{760\text{ torr}}\right)^3 = 0.0196$$

$$K_{P450} = \left(\frac{4}{27}\right)\left(\frac{810\text{ torr}}{760\text{ torr}}\right)^3 = 0.1794$$

(b) $$\Delta H° = \frac{RT_1T_2}{T_2 - T_1}\ln\frac{K_2}{K_1}$$

$$= \frac{(1.987\text{ cal K}^{-1}\text{ mol}^{-1})(693\text{ K})(723\text{ K})}{30\text{ K}}\ln\frac{0.1794}{0.0196}$$

$= 73,500$ cal mol^{-1} for the reaction as written

$= 36,750$ cal mol^{-1} of HgO (s)

4.31 (a) Calculate the equilibrium constants of the following reactions at 298 K:

$$C(\text{graphite}) + 2H_2(g) = CH_4(g)$$

$$2C(\text{graphite}) + 3H_2(g) = C_2H_6(g)$$

(b) Assuming $\Delta H°$ is independent of temperature at what temperature will these reactions have equilibrium constants of unity?

SOLUTION:

(a) $C\,(\text{graphite}) + 2H_2\,(g) = CH_4\,(g)$

$$\Delta G^\circ_{298} = -12,130 = -RT \ln K$$

$$= -(1.987 \text{ cal K}^{-1} \text{ mol}^{-1})(298.15 \text{ K}) \ln K_p$$

$$K_p = 7.80 \times 10^8$$

$2C\,(\text{graphite}) + 3H_2\,(g) = C_2H_6\,(g)$

$$\Delta G^\circ_{298} = -7860 = -RT \ln K$$

$$= -(1.987 \text{ cal K}^{-1} \text{ mol}^{-1})(298.15 \text{ K}) \ln K_p$$

$$K_p = 5.78 \times 10^5$$

(b) For the first reaction, $\Delta H = -17.88$ kcal mol^{-1}

$$\ln \frac{K_2}{K_1} = \frac{\Delta H^\circ (T_2 - T_1)}{R T_1 T_2}$$

$$\ln \frac{1}{7.80 \times 10^8} = \frac{-(17,880 \text{ cal mol}^{-1})(T_2 - 298 \text{ K})}{(1.987 \text{ cal K}^{-1} \text{ mol}^{-1})(298 \text{ K}) T_2}$$

$$T_2 = 925 \text{ K}$$

For the second reaction, $\Delta H^\circ = -20.24$ Kcal mol^{-1}

$$\ln \frac{1}{5.78 \times 10^5} = \frac{-(20,240 \text{ cal mol}^{-1})(T_2 - 298 \text{ K})}{(1.987 \text{ cal K}^{-1} \text{ mol}^{-1})(298 \text{ K}) T_2}$$

$$T_2 = 487 \text{ K}$$

4.32 Using values for the ion product for water at 25° C from Table 18.3 calculate ΔH° for

$$H_2O\,(l) = H^+\,(aq) + OH^-\,(aq)$$

and compare it with the value calculated from Table A.1.

SOLUTION:

Using values at 10°C and 40° from Table 18.3,

$$\Delta H^\circ = \frac{RT_1T_2}{T_2-T_1} \ln \frac{K_2}{K_1}$$

$$= \frac{(1.987 \text{ cal K}^{-1} \text{ mol}^{-1})(283.1\text{K})(313.1\text{K})}{30 \text{ K}} \ln \frac{2.917}{0.292}$$

$$= 13,500 \text{ cal mol}^{-1}$$

Using Table A.1

$$\Delta H^\circ = -54.970 + 68.313 = 13.345 \text{ kcal mol}^{-1}$$

4.33 The following reaction is nonspontaneous at room temperature and endothermic:

$$3C\,(\text{graphite}) + 2H_2O(g) = CH_4(g) + 2CO(g)$$

As the temperature is raised the equilibrium constant will become equal to unity at some point. Estimate this temperature using data from Table A.2

SOLUTION:

AT 1000K

$$\Delta G^\circ = 4.625 - 2(47.859) + 2(46.040)$$

$$= 0.987 \text{ kcal mol}^{-1}$$

$$= -(1.987 \times 10^{-3}\ \text{kcal K}^{-1}\ \text{mol}^{-1})(1000\ \text{K}) \ln K_p$$

$$K_p = 0.609$$

$$\Delta H° = -21.482 - 2(26.771) + 2(59.246)$$

$$= +43.47\ \text{kcal mol}^{-1}$$

$$\ln \frac{K_2}{K_1} = \frac{\Delta H°}{R}\left(\frac{1}{T_1} - \frac{1}{T_2}\right)$$

$$\ln \frac{1}{0.609} = \frac{43,470\ \text{cal mol}^{-1}}{1.987\ \text{cal K}^{-1}\ \text{mol}^{-1}}\left(\frac{1}{1000\text{K}} - \frac{1}{T_2}\right)$$

$$T_2 = \frac{1}{\dfrac{1}{1000\text{K}} - \dfrac{1.987}{43,470} \ln \dfrac{1}{0.609}}$$

$$= 1023\ \text{K}$$

4.34 Calculate the degree of dissociation of $H_2O(g)$ into $H_2(g)$ and $O_2(g)$ at 2000K and 1atm. (Since the degree of dissociation is small, the calculation may be simplified by assuming that $P_{H_2O} = 1$atm.)

SOLUTION:

$$H_2O(g) = H_2(g) + \tfrac{1}{2} O_2(g)$$

$$\Delta G°(2000\ \text{K}) = 32.401\ \text{kcal mol}^{-1}$$

$$= -RT \ln K_p$$

$$K_p = e^{-32,401/(1.987)(2000)}$$

$$= 2.88 \times 10^{-4}$$

$$= \frac{(P_{H_2}/P^\circ)(P_{O_2}/P^\circ)^{1/2}}{(P_{H_2O}/P^\circ)}$$

$$= \frac{P_{H_2}}{P^\circ}\left(\frac{P_{H_2}}{2P^\circ}\right)^{1/2} = \frac{P_{H_2}^{\;3/2}}{\sqrt{2}}$$

$$\frac{P_{H_2}}{P^\circ} = (\sqrt{2}\ 2.88 \times 10^{-4})^{2/3}$$

$$P = 0.0055 \text{ atm}$$

$$\alpha = \frac{0.0055 \text{ atm}}{1 \text{atm}} = 0.0055$$

4.36 The equilibrium constant for the association of benzoic acid to a dimer in dilute benzene solutions is as follows at 43.9°C:

$$2\, C_6H_5COOH = (C_6H_5COOH)_2 \quad K_c = 2.7 \times 10^2$$

Molar concentrations are used in expressing the equilibrium constant. Calculate ΔG°_c and state its meaning.

SOLUTION:

$$\Delta G^\circ = -RT \ln K_c$$

$$= -(1.987 \text{ cal K}^{-1} \text{ mol}^{-1})(317.1 \text{ K}) \ln (2.7 \times 10^2)$$

$$= -3530 \text{ cal mol}^{-1}$$

This is the decrease in Gibbs energy when 2 mol of monomer at unit activity on the molar scale is converted to 1 mol of dimer at unity activity on the molar scale.

4.37 $\Delta G° > 4110$ cal mol^{-1}

4.38 7.1 atm

4.39 64 %

4.40 0.696

4.41 (a) 19.45 atm (b) 35.1 atm

4.42 (a) 2.21×10^{-5} (b) 5.41×10^{-4}

4.43 10.5

4.44 (a) 5×10^{-3} atm (b) 1.79 % (c) 0.0221 atm

4.45 7.47 atm

4.46 1.87×10^{-4}, 1.64×10^{-4}, 1.90×10^{-4}, 0.364×10^{-4}

As the pressure is increased, the gases behave more imperfectly.

4.47 3.69 atm

4.48 (a) No (b) 0.285

4.49 6.85

4.50 (a) 2.48, (b) 0.540 mol

4.51 $\alpha = 0.800$ $K_p = 1.78$

4.52 (a) 4054 cal mol^{-1} (b) 0.785 atm

4.53 (a) 9.63 g (b) 1.199, 0.032, 1.438 atm

4.54 $k = 3^{3/2}/4 K_p = 0.0166$

4.55 (a) 9.75 g (b) 19.75 g

4.56 3.31×10^{-16} atm of O_2 is required to oxidize Ni, but the partial pressure of O_2 from partially dissociated H_2O in the presence of 3 mole percent hydrogen is only 9.7×10^{-18} atm.

4.57 Dissociation into atoms and solubility proportional to the partial pressure of atoms.

4.58 5.50×10^{-7} atm

4.59 1.83 kcal mol^{-1} 1.44 kcal mol^{-1} 0.088

4.60 Fe_2O_3

4.61 1.81×10^{-2}

4.62 2.99×10^{-3} 0.0273 0.0861

4.63 $K_p = 0.081$ 0.363

4.64 108 kcal mol^{-1}

4.65 (a)

$t/°C$	25	45	65
α	0.1852	0.3775	0.6284
K_p	0.1432	0.6649	2.610

(b) 14.5 kcal mol^{-1} (c) 0.320 (d) 0.371

4.66 137°C

4.67 36.6 kcal mol^{-1}

4.68 1.92×10^{-13} atm

4.69 (a) 6.88×10^{-24}, 4.79×10^{-43} (b) 1005, 1359 K

4.70 2.14×10^{-4} atm

4.71 − 77.5 kcal mol^{-1}

4.72 541

4.73 16.6

CHAPTER 5. Electrochemical Cells

5.1 How much work is required to bring two protons from an infinite distance of separation to 0.1 nm? Calculate the answer in joules using the protonic charge 1.602×10^{-19} C. What is the work in kcal mol^{-1} for a mole of proton pairs?

SOLUTION:

$$\text{Potential } \varphi = \frac{Q_2}{4\pi\varepsilon_0 r}$$

$$= \frac{(1.602 \times 10^{-19}\ C)(0.89875 \times 10^{10}\ N\,m^2\,C^{-2})}{10^{-10}\ m}$$

$$= 14.398\ J\,C^{-1}$$

$$\text{Work} = Q_1\varphi = (1.602 \times 10^{-19}\ C)(14.398\ J\,C^{-1})$$

$$= 2.307 \times 10^{-18}\ J$$

$$= \frac{(2.307 \times 10^{-18}\ J)(6.022 \times 10^{23}\ mol^{-1})}{4.184\ J\,cal^{-1}}$$

$$= 332.0\ kcal\ mol^{-1}$$

5.2 A small dry battery of zinc and ammonium chloride weighing 85 g will operate continuously through a 4 Ω resistance for 450 min before its voltage falls below 0.75 V. The initial voltage is 1.60 and the effective voltage over the whole life of the battery is taken to be 1.00. Theoretically, how many miles above earth could this battery be raised by the energy delivered under these conditions?

SOLUTION:

$$I = \frac{E}{R} = \frac{1V}{4\Omega} = 0.25 \text{ A}$$

$$\text{Power} = I^2R = (0.25\text{A})^2(4\Omega) = 0.25 \text{ W}$$

$$\text{Work} = (0.25 \text{ W})(450 \times 60 \text{ s}) = 6.75 \times 10^3 \text{ J}$$

$$= (0.085 \text{ Kg})(9.80 \text{ ms}^{-2})h$$

$$h = \frac{6.75 \times 10^3 \text{ J}}{(0.085 \text{ Kg})(9.80 \text{ ms}^{-2})}$$

$$= 8103 \text{ m}$$

$$= \frac{(8103 \times 10^5 \text{ cm})}{(2.54 \text{cm in}^{-1})(12 \text{ in ft}^{-1})(5280 \text{ ft mile}^{-1})}$$

$$= 5.04 \text{ miles}$$

5.3 The voltage of the cell

$$Pb \mid PbSO_4 \mid Na_2SO_4 \cdot 10\,H_2O(\text{sat}) \mid Hg_2SO_4 \mid Hg$$

is 0.9647 at 25°C. The temperature coefficient is 1.74×10^{-4} V K^{-1}.

(a) What is the cell reaction?

(b) What are the values of ΔG, ΔS, and ΔH?

SOLUTION

(a) $Pb(s) + Hg_2SO_4(s) = PbSO_4(s) + 2\,Hg(\ell)$

(b) $\Delta G = -zFE = -(2)(23{,}060 \text{ cal mol}^{-1}\text{V}^{-1})(0.9647 \text{ V})$

$$= -44{,}492 \text{ cal mol}^{-1}$$

$$\Delta S = zF\left(\frac{\partial E}{\partial T}\right)_p$$

$$= (2)(23{,}060 \text{ cal mol}^{-1}\text{V}^{-1})(1.74 \times 10^{-4} \text{ VK}^{-1})$$

$$= 8.02 \text{ cal K}^{-1} \text{ mol}^{-1}$$

$$\Delta H = -zFE + zFT\frac{\partial E}{\partial T}$$

$$= -2(23{,}060 \text{ cal mol}^{-1}\text{V}^{-1})(0.9647 \text{ V})$$

$$+ 2(23{,}060 \text{ cal mol}^{-1}\text{V}^{-1})(298.15 \text{ K})(1.74 \times 10^{-4} \text{ VK}^{-1})$$

$$= -42{,}100 \text{ cal mol}^{-1}$$

5.4 Derive an expression for the activity of a 1-2 electrolyte (like Na_2SO_4) in terms of the mean ionic activity coefficient and the molality.

SOLUTION:

$$a_{Na_2SO_4} = a^2_{Na^+} \cdot a_{SO_4^=} \qquad \gamma_{Na^+} = \frac{a_{Na^+}}{2m}$$

$$= (2m\gamma_{Na^+})^2(\gamma_{SO_4^=}m) \qquad \gamma_{SO_4^=} = \frac{a_{SO_4^=}}{m}$$

$$= 4m^3\gamma_\pm^3$$

where $\gamma_\pm = (\gamma_{Na^+}^2 \cdot \gamma_{SO_4^=})^{1/3}$

5.6 For 0.002 mol kg^{-1} $CaCl_2$ at 25°C use the Debye-Huckel limiting law to calculate the activity coefficients of Ca^{++} and Cl^-. What is the mean ionic activity coefficient for the electrolyte?

SOLUTION:

$$I = \frac{1}{2}(0.002 \times 2^2 + 0.004 \times 1^2) = 0.006$$

$$\log \gamma_i = -0.509\, z_i^2\, I^{1/2}$$

$$\log \gamma_{Ca^{2+}} = -0.509\,(2^2)(0.006)^{1/2}$$

$$\gamma_{Ca^{2+}} = 0.695$$

$$\log \gamma_{Cl^-} = -0.509\,(1^2)(0.006)^{1/2}$$

$$\gamma_{Cl^-} = 0.913$$

$$\gamma_\pm = (\gamma_+\, \gamma_-^{\,2})^{1/3} = (0.695 \times 0.913^2)^{1/3}$$

$$= 0.834$$

5.7 The cell $Pt|H_2\,(1\,atm)|HBr\,(m)|AgBr|Ag$ has been studied by H. S. Harned, A. S. Keston, and J. G Donelson [J. Am. Chem. Soc., 58, 989 (1936)]. The following table gives the emf's obtained at 25° C:

m	0.01	0.02	0.05	0.10
E	0.3127	0.2786	0.2340	0.2005

Calculate (a) E° and (b) the activity coefficient for a 0.10 mol kg^{-1} solution of hydrogen bromide.

SOLUTION:

(a) Plot $E + 0.1183 \log m - 0.0602 \sqrt{m}$ versus m, The intercept at m = 0 is E° = 0.0710 V.

(b) $$E = 0.0710\,V - (0.05915\,V) \log \gamma_\pm^2\, m^2$$

$$0.2005\,V = 0.0710\,V - (0.1183\,V) \log \gamma_\pm - (0.1183\,V) \log 0.1$$

$$\log \gamma_\pm = -\frac{0.2005 - 0.1183 - 0.0710}{0.1183}$$

$$\gamma_{\pm} = 0.804$$

5.8 Given the cell at 25° C

$$Pb \mid Pb^{2+} (a=1) \parallel Ag^{+} (a=1) \mid Ag$$

(a) calculate the voltage; (b) write the cell reaction; and (c) calculate the Gibbs energy change. (d) Which electrode is positive?

SOLUTION:

(a) $E^\circ = 0.126 + 0.7991 = 0.925\ V$

(b) $Pb = Pb^{2+} + 2e \qquad -0.126\ V$

$$2Ag^{+} + 2e = 2Ag \quad +0.799V$$

$$Pb\,(s) + 2\,Ag^{+}\,(a=1) = Pb^{2+}\,(a=1) + 2\,Ag\,(s) + 0.925$$

or

$$\tfrac{1}{2}\,Pb\,(s) + Ag^{+}\,(a=1) = \tfrac{1}{2}\,Pb^{2+}\,(a=1) + Ag\,(s)$$

(c) $\Delta G^\circ = -zFE^\circ = -(1)\,(23{,}060\ cal\ mol^{-1}V^{-1})(0.925\ V)$

$$= -21{,}330\ cal\ mol^{-1}$$

(d) Silver electrode

5.9 (a) Calculate the voltage of the following cell at 25° C.

$$Zn \mid Zn^{2+} (a = 0.0004) \parallel Cd^{2+} (a = 0.2) \mid Cd$$

(b) Write the cell reaction. (c) Calculate the value of the Gibbs energy change involved in the reaction.

SOLUTION:

(a) $Zn\,(s) = Zn^{2+}\,(a = 0.0004) + 2e \qquad E^\circ = -0.763\ V$

$$\begin{array}{ll} Cd^{2+}(a=0.2) + 2e = Cd(s) & E° = -0.403\ V \\ \hline Zn(s) + Cd^{2+}(a=0.2) = Cd(s) + Zn^{2+}(a=0.0004) & \end{array}$$

$$E° = -0.403 - (-0.763) = 0.360\text{V}$$

$$E = E° - \frac{RT}{zF} \ln \frac{a_{Zn^{2+}}}{a_{Cd^{2+}}}$$

$$= 0.360 - \frac{0.05916}{2} \log \frac{0.0004}{0.2}$$

$$= 0.440\ \text{V}$$

(b) See (a)

(c) $\Delta G = -zFE = -2\ (23{,}060\ \text{cal mol}^{-1}\ \text{V}^{-1})(0.440\text{V})$

$$= -20{,}300\ \text{cal mol}^{-1}$$

5.11 (a) Calculate the equilibrium constant at 25° C for the reaction

$$Fe^{2+} + Ag^{+} = Ag + Fe^{3+}$$

(b) Calculate the concentration of silver ion at equilibrium (assuming that concentrations may be substituted for activities) for an experiment in which an excess of finely divided metallic silver is added to a 0.05 mol kg^{-1} solution of ferric nitrate.

SOLUTION:

(a) $Fe^{2+} = Fe^{3+} + e$ $\qquad E° = 0.771\text{V}$

$$\begin{array}{ll} Ag^{+} + e = Ag(s) & E° = 0.7991\text{V} \\ \hline Fe^{2+} + Ag^{+} = Fe^{2+} + Ag(s) & E° = 0.0281\text{V} \end{array}$$

$$\log K = \frac{z\ FE°}{2.303\ RT} = \frac{0.0281\ \text{V}}{0.05916\text{V}} = 0.475$$

$K = 3.00$

(Strictly speaking an equilibrium constant does not have units, but since we are assuming that molar concentrations may be substituted for activities, it is well to remind ourselves of the units this gives K).

$$(b)\quad K = \frac{(Fe^{3+})}{(Fe^{2+})(Ag^{+})} = \frac{0.05 - X}{X^2} = 3.0$$

$$X = 0.044 \text{ mol kg}^{-1}$$

5.12 Devise an electromotive force cell for which the cell reaction is

$$AgBr(s) = Ag^{+} + Br^{-}$$

Calculate the equilibrium constant (usually called the solubility product) for this reaction at 25°C.

SOLUTION:

$$Ag\,|\,Ag^{+}\,\|\,Br^{-}\,|\,AgBr\,|\,Ag$$

$$Ag = Ag^{+} + e \qquad E° = 0.7991\ V$$

$$AgBr + e = Ag + Br^{-} \qquad E° = 0.095\ V$$

$$AgBr = Ag^{+} + Br^{-} \qquad E° = -0.701\ V$$

$$\log K = \frac{E°}{0.05916} = \frac{-0.701 V}{0.05916 V} = -11.85$$

$$K = 10^{-11.85} = m^2 \gamma_{\pm}^2$$

5.13 What are the equilibrium constants for the following reactions at 25°C?

(a) $H^{+}(aq) + Li(c) = Li^{+}(aq) + \frac{1}{2}H_2(g)$

(b) $2H^+ (aq) + Pb(c) = Pb^{2+} (aq) + H_2 (g)$

(c) $3H^+ (aq) + Au(c) = Au^{3+} (aq) + \frac{3}{2} H_2 (g)$

SOLUTION:

(a)
$$\begin{array}{ll} Li = Li^+ + e & -3.045\ V \\ H^+ + e = \frac{1}{2} H_2 & 0 \\ \hline H^+ + Li = Li^+ + \frac{1}{2} H_2 & E° = 3.045\ V \end{array}$$

$$K = e^{zFE°/RT}$$

$$= e^{(23{,}060\ cal\ mol^{-1} V^{-1})(3.045\ V)/(1.987\ cal\ K^{-1} mol^{-1}) \times (298\ K)}$$

$$= 3.1 \times 10^{51}$$

(b)
$$\begin{array}{ll} Pb = Pb^{2+} + 2e & -0.126\ V \\ 2H^+ + 2e = H_2 & 0 \\ \hline 2H^+ + Pb = Pb^{2+} + H_2 & E° = 0.126\ V \end{array}$$

$$K = e^{2(23{,}060\ cal\ mol^{-1} V^{-1})(0.126 V)/1.987\ cal\ K^{-1} mol^{-1}) \times (298 K)}$$

$$= 1.8 \times 10^4$$

(c)
$$\begin{array}{ll} Au = Au^{3+} + 3e & 1.50\ V \\ 3H^+ + 3e = \frac{3}{2} H_2 & 0 \\ \hline 3H^+ + Au = Au^{3+} + \frac{3}{2} H_2 & E° = -1.50\ V \end{array}$$

$$K = e^{3(23{,}060\ cal\ mol^{-1} V^{-1})(-1.50V)/(1.987\ cal\ K^{-1} mol^{-1}) \times (298\ K)}$$

$= 7.8 \times 10^{-77}$

5.14 For the reaction

$$\tfrac{1}{2} Cu(s) + \tfrac{1}{2} Cl_2(g) = \tfrac{1}{2} Cu^{2+} + Cl^-$$

at 25° C calculate (a) the standard Gibbs energy change, (b) the equilibrium constant, and (c) the standard electromotive force of the cell in which this reaction occurs.

SOLUTION:

(a) Oxidation $\tfrac{1}{2} Cu(s) = \tfrac{1}{2} Cu^{2+} + e$

Reduction $\tfrac{1}{2} Cl_2(g) + e = Cl^-$

Since the oxidation reaction is to occur at the left electrode and the reduction at the right electrode, the cell is written

$$Cu \,|\, Cu^{2+} \,\|\, Cl^- \,|\, Cl_2(g) \,|\, Pt$$

$$E° = E°_{Cl^-|Cl_2|Pt} - E°_{Cu^{2+}|Cu} = 1.360 - 0.337 = 1.023 V$$

$$\Delta G° = -zFE° = -(23{,}060 \text{ cal mol}^{-1} V^{-1})(1.023 V)$$

$$= 23{,}590 \text{ cal mol}^{-1}$$

(b) $K = e^{(23{,}590 \text{ cal mol}^{-1})/(1.987 \text{ cal K}^{-1} \text{mol}^{-1})(298.15 K)}$

$= 1.97 \times 10^{17}$

(c) See (a)

Alternatively this problem may be done using data in Table A.1.

(a) $\Delta G° = \tfrac{1}{2}(15.66) - 3.372 = -23.54$ kcal mol^{-1}

(b) $K = e^{(23,540 \text{ cal mol}^{-1})/(1.987 \text{ cal K}^{-1} \text{ mol}^{-1})(298.15 \text{K})}$

$= 1.80 \times 10^{17}$

(c) $E° = \frac{-\Delta G°}{zF} = \frac{23,540 \text{ cal mol}^{-1}}{23,060 \text{ cal mol}^{-1} \text{ V}^{-1}}$

$= 1.021 \text{ V}$

5.16 (a) Calculate the equilibrium constant at 25°C for the reaction

$$Sn^{4+} + 2Ti^{3+} = 2Ti^{4+} + Sn^{2+}$$

(b) When 0.01 mol of Sn^{2+} ion is added to 1.0 mol of Ti^{4+} ion in 1000 g of water, what will be the concentration of Ti^{3+} ions (if it is assumed for the calculation that the activities are equal to the concentrations)?

SOLUTION:

(a) $2Ti^{3+} = 2Ti^{4+} + 2e \qquad E° = 0.04 \text{V}$

$Sn^{4+} + 2e = Sn^{2+} \qquad E° = 0.15 \text{ V}$

$Sn^{4+} + 2Ti^{3+} = 2Ti^{4+} Sn^{2+} \qquad E° = 0.11 \text{ V}$

$K = e^{zFE°/RT} = e^{2(23,060)(0.11)/(1.987)(298)}$

$= 5.26 \times 10^3$

(b) $Sn^{4+} + 2Ti^{3+} = 2Ti^{4+} + Sn^{2+}$

$\quad X \qquad 2X \qquad 1.0-2X \qquad 0.01-X$

$K = (1.0-2x)^2 \ (0.01-x)/4x^3 = 5.26 \times 10^3$

$(Ti^{3+}) = 2x = 1.16 \times 10^{-2} \text{ mol kg}^{-1}$

5.17 (a) Diagram the cell that corresponds to the reaction $Ag^+ + Cl^- = AgCl$. Calculate at 25° C (b) E°, (c) ΔG°, (d) K.

SOLUTION:

(a) $Ag | AgCl | Cl^- || Ag^+ | Ag$

(b)
$$Ag + Cl^- = AgCl + e^- \qquad E° = -0.2224\text{V}$$
$$Ag^+ + e = Ag \qquad E° = 0.7991\text{V}$$
$$Ag^+ + Cl^- = AgCl \qquad E° = 0.5767\text{V}$$

(c)
$$\Delta G° = -zFE° = -(23{,}060 \text{ cal mol}^{-1}\text{V}^{-1})(0.5767\text{V})$$
$$= -13{,}300 \text{ cal mol}^{-1}$$

(d)
$$K = e^{-\Delta G°/RT}$$
$$= e^{13{,}300/(1.987)(298)}$$
$$= 5.6 \times 10^9$$

5.18 Show that for the concentration cell

$$X | X^-(a_1) || X^-(a_2) | X$$

where X^- is a negative ion, the equation for the electromotive force of the cell is

$$E = -\frac{RT}{F} \ln \frac{a_2}{a_1}$$

SOLUTION:

$$X + e = X^-(a_2) \qquad E = E° - \frac{RT}{F} \ln a_2$$

$$X^-(a_1) = X + e \qquad E = -\left(E° - \frac{RT}{F} \ln a_1\right)$$

$$X^-(a_1) = X^-(a_2) \qquad E = -\frac{RT}{F} \ln \frac{a_2}{a_1}$$

5.19 Given the cell at 25° C

$$Pt \mid Cl_2 \mid Cl^- \ (a = 0.1) \parallel Cl^- (a = 0.001) \mid Cl_2 \mid Pt$$

(a) Write the cell reaction. (b) Which electrode is negative? What is the voltage of the cell? (d) Is the reaction spontaneous?

SOLUTION:

(a)

$$Cl^- (a = 0.1) = \tfrac{1}{2}\ Cl_2 + e$$

$$\tfrac{1}{2}\ Cl_2 + e = Cl^- \ (a = 0.001)$$

$$Cl^- \ (a = 0.1) = Cl^- \ (a = 0.001)$$

(b) The left electrode is negative. If we wish we can think of this resulting from more frequent collisions with halogen ions.

(c) $E = -\frac{RT}{zF} \ln Q = -0.05916 \log \frac{0.001}{0.01}$

$= 0.11832$ V

(d) Yes. The electromotive force of the cell tells us this, but we can also note that dilution is a spontaneous process.

5.21 Ammonia may be used as the anodic reactant in a fuel cell. The reactions occuring at the electrodes are

$$NH_3(g) + 3OH^-(aq) = \tfrac{1}{2} N_2(g) + 3H_2O(\ell) + 3e$$

$$O_2(g) + 2H_2O(\ell) + 4e = 4OH^-(aq)$$

What is the electromotive force of this fuel cell at 25°C?

SOLUTION:

$$4NH_3 + 12OH^- = 2N_2 + 12H_2O + 12e$$

$$3O_2 + 6H_2O + 12e = 12OH^-$$

$$4NH_3 + 3O_2 = 2N_2 + 6H_2O$$

Using Table A.1

$$\Delta G° = 6(-56.69) - 4(-3.94) = -324.38 \text{ Kcal mol}^{-1}$$

$$= -zFE°$$

$$E° = \frac{-324,380 \text{ cal mol}^{-1}}{-(12)(23,060 \text{ cal mol}^{-1} V^{-1})}$$

$$= 1.1722 \text{ V}$$

5.22 Calculate the electromotive force of a methane$-O_2$ fuel cell at 25° C.

SOLUTION

$$CH_4(g) + 2H_2O(\ell) = CO_2(g) + 8H^+ + 8e$$

$$2O_2(g) + 8e + 8H^+ = 4H_2O(\ell)$$

$$CH_4(g) + 2O_2(g) = CO_2(g) + 2H_2O(\ell)$$

$$\Delta G° = -94.254 + 2(-56.687) - (-12.13)$$

$$= -195.50 \text{ kcal mol}^{-1}$$

$$= -zFE°$$

$$E° = \frac{-195.50 \text{ kcal mol}^{-1}}{8(23.060 \text{ kcal mol}^{-1} \text{V}^{-1})}$$

$$= 1.0597 \text{V}$$

5.23 Calculate the electromotive force of

$$\text{Li}(\ell) \mid \text{LiCl}(\ell) \mid \text{Cl}_2(g)$$

at 900 K for $P_{Cl_2} = 1$ atm. This high temperature battery is attractive because of its high electromotive force and low relative atomic masses. Lithium chloride melts at 883 K and lithium at 453.69K. (The ΔG_f° for LiCl(ℓ) at 900 K in JANAF Thermochemical Tables is -80.100 kcal mol^{-1}.)

SOLUTION:

$$\text{Li}(\ell) + \tfrac{1}{2}\text{Cl}_2(g) = \text{LiCl}(\ell)$$

$$\Delta G° = -80{,}100 \text{ cal mol}^{-1} = -zFE°$$

$$E° = \frac{-80{,}100 \text{ cal mol}^{-1}}{-23{,}060 \text{ cal mol}^{-1} \text{V}^{-1}}$$

$$= 3.474 \text{V}$$

5.24 66.4 kcal mol^{-1}

5.25 110.2, 11.02, 1.38 kcal mol^{-1}

5.26 (a) $\text{Zn} + 2\text{AgCl} = 2\text{Ag} + \text{ZnCl}_2$ (0.555 mol kg^{-1})

(b) -46.7 kcal mol^{-1} (c) -18.5 cal K^{-1} mol^{-1}

(d) −52.21 kcal mol^{-1}

5.27 0.905

5.28 (a) $[(a_+)(a_-)]^{1/2}/m$

(b) $[(a_+)(a_-)^3]^{1/4}/3^{3/4}\ m$

(c) $[(a_+)(a_-)]^{1/2}/m$

5.29 (a) 0.24, (b) 0.08, (c) 0.06 mol kg^{-1}

5.30 (a) Na | NaOH (m) | H_2, Pt

At 25°C $E = E^\circ - 0.0591 \log (m^2 \gamma_\pm^2 P_{H_2}^{1/2})$

(b) Pt | H_2 | H_2SO_4 (m) | Ag_2SO_4 | Ag

At 25° C $E = E^\circ - 0.0296 \log (4m^3 \gamma_\pm^3 P_{H_2}^{-1})$

5.31 $pK = 1.018\sqrt{I} + \log\frac{m_1}{m_2} + \frac{(E - E_\circ)F}{2.303\ RT} + \log m_3$

5.32 171.4 kcal mol^{-1} 3.717V

5.33 5.92 V

5.34 (a) $Cd(s) + I_2(s) = Cd^{2+}(a=1) + 2I^-(a=1)$

(b) 0.939V (c) −43,300 cal mol^{-1}

(d) The iodine electrode is positive.

5.35 (a) Pt | Br_2 (l) | Br^- (a=1) ‖ Cl^- (a=1) | Cl_2 (g) | Pt

(b) 0.2943 V (c) −6787 cal mol^{-1}

5.36 (a) 0.403 V (b) −9295 cal mol^{-1}

5.37 (a) 1.41 V

(b) $Ti^{3+}(a = 0.3) + Ce^{4+}(a = 0.002) = Ti^{2+}(a = 0.5) + Ce^{3+}(a = 0.7)$

(c) −32.5 kcal mol^{-1} (d) 1.57 V

(e) −36.2 cal mol^{-1} (f) 3.46×10^{26}

5.38 5.64 V

5.39 3.56×10^{3} watt-hr kg^{-1}

5.40 1.30

5.41 2×10^{-8} mol L^{-1}

5.42 0.0296 V

5.43 (a) The electrode with the higher percent thallium is negative.

(b) −348 cal mol^{-1} (c) 0.030462 V

5.44 (a) 2.62 (b) 2.4×10^{-3}

5.45 1.21 V

5.46 0.951 V

5.47 (a) 1.229 V

(b) 1.229 V

5.48 0.60

CHAPTER 6 Equilibria of Biochemical Reactions

6.1 Calculate the pH of (a) a 0.1 mol L^{-1} solution of n-butyric acid, and (b) a solution containing 0.05 mol L^{-1} butyric acid and 0.05 mol L^{-1} sodium butyrate. Using these data, sketch the titration curve for 0.1 mol L^{-1} butyric acid that is titrated with a strong base so concentrated that the volume of the solution may be considered to remain constant ($K = 1.48 \times 10^{-5}$ at 25°C)

SOLUTION:

(a) $K = 1.48 \times 10^{-5} = \dfrac{x^2}{0.1}$

$x = (H^+) = 1.22 \times 10^{-3}$ and $pH = -\log(1.22 \times 10^{-3}) = 2.91$

(b) $pH = pK + \log \dfrac{C_s}{C_a}$

$= 4.83 + \log \dfrac{0.05}{0.05} = 4.83$

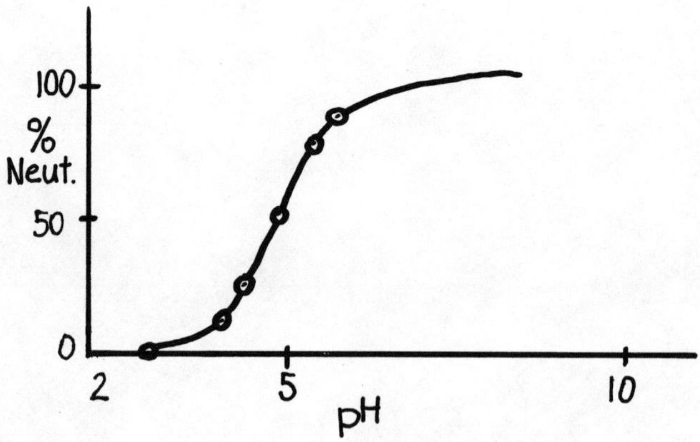

6.2 A buffer contains 0.01 mol of lactic acid ($pK = 3.60$) and 0.05 mol of sodium lactate per liter. (a) Calculate the pH of this buffer. (b) Five milliliters of 0.5 mol L^{-1} hydrochloric acid is added to a liter of the buffer. Calculate the change in pH. (c) Calculate the pH change to be expected if this quanity of acid is added to 1 liter of a solution of a strong acid of the same initial pH.

SOLUTION:

(a) $$pH = pK + \log \frac{C_s}{C_a}$$

$$= 3.60 + \log \frac{0.05}{0.01} = 4.30$$

(b) $(0.005 \text{ liter})(0.5 \text{ mol liter}^{-1}) = 2.5 \times 10^{-3} \text{ mol}$

$$pH = 3.60 + \log \frac{0.05 - 0.0025}{0.01 + 0.0025} = 4.18$$

$$\Delta pH = 4.18 - 430 = -0.12$$

(c) Strong acid of pH 4.30

$$(H^+) = \text{antilog } (-4.30) = 5.0 \times 10^{-5}$$

After addition of HCl,

$$(H^+) = 5.0 \times 10^{-5} + 2.5 \times 10^{-3}$$

$$= 2.55 \times 10^{-3}$$

$$pH = -\log (2.55 \times 10^{-3}$$

$$= 2.54$$

$$\Delta pH = 2.54 - 4.30$$

$$= 1.76$$

6.3 Calculate the number of moles per liter of Na_2HPO_4 and NaH_2PO_4 which should be used to prepare 0.10 mol L^{-1} ionic strength (equation 5.22) buffer of pH 7.30. At this ionic strength the second pK of phosphoric acid may be taken as 6.84.

SOLUTION:

$$pH = pK + \log \frac{(HPO_4^{2-})}{(H_2PO_4^{1-})}$$

$$7.30 = 6.84 + \log \frac{(HPO_4^{2-})}{(H_2PO_4^{1-})}$$

$$(HPO_4^{2-}) = 2.88\,(H_2PO_4^{1-})$$

$$I = 0.10 = \tfrac{1}{2}\left[(H_2PO_4^{-1}) \times 1^2 + (HPO_4^{2-}) \times 2^2 + (Na^+) \times 1^2\right]$$

$$= \tfrac{1}{2}\left[(H_2PO_4^{1-}) + (4)(2.88)(H_2PO_4^{1-}) + (1 + 2 \times 2.88) \times (H_2PO_4^{1-})\right]$$

$$(H_2PO_4^{1-}) = 0.104 \text{ mol } L^{-1}$$

$$(H_2PO_4^{2-}) = 0.299 \text{ mol } L^{-1}$$

6.5 According to Table A.1 what are the values of ΔG°, ΔH°, ΔS° at 298K for

$$H_2O\,(\ell) = H^+(aq) + OH^-(aq)$$

Show that the same value of ΔS° is obtained from ΔG° and ΔH° as by using $\Delta S^\circ = \sum \nu_i S_i^\circ$.

SOLUTION

$\Delta G° = -37{,}594 + 56{,}687 = 19{,}093$ cal mol^{-1}

$\Delta H° = -54{,}970 + 68{,}315 = 13{,}345$ cal mol^{-1}

$\Delta S° = -2.57 - 16.71 = -19.28$ cal K^{-1} mol^{-1}

or

$$\Delta S° = \frac{\Delta H° - \Delta G°}{T} = \frac{(13{,}345 - 19{,}093) \text{ cal mol}^{-1}}{298.15 \text{ K}}$$

$$= -19.28 \text{ cal K}^{-1} \text{ mol}^{-1}$$

6.6 Given the following thermodynamic quantities calculate the pH of the midpoint of titration of the first acid group of carbonic acid at 25°C.

	$\Delta H°$ Kcal mol^{-1}	$\Delta S°$ cal K^{-1} mol^{-1}
$CO_2 + H_2O = H_2CO_3$	1.13	-8
$H_2CO_3 = H^+ + HCO_3$	1.01	-14

SOLUTION:

$$K_\alpha = \frac{(H_2CO_3)}{(CO_2)} = 2.36 \times 10^{-3} \text{ since } \Delta G° = 3515 \text{ cal mol}^{-1}$$

$$K_\beta = \frac{(H^+)(HCO_3^-)}{(H_2CO_3)} = 1.55 \times 10^{-4} \text{ since } \Delta G° = 5180 \text{ cal mol}^{-1}$$

$$K_1 = \frac{(H^+)(HCO_3^-)}{(H_2CO_3) + (CO_2)} = \frac{(H^+)(HCO_3^-)}{(H_2CO_3)\left(1 + \frac{1}{K_\alpha}\right)}$$

$$= \frac{K_\beta}{1 + \frac{1}{K_\alpha}} = 4.07 \times 10^{-7}$$

$pK_1 = 6.39$ This is the pH of the middle of the titration curve.

6.7 Estimate pK_1 and pK_2 for H_3PO_4 at 25°C and 0.1 ionic strength. The values at zero ionic strength are

$$pK_1 = 2.148$$

$$pK_2 = 7.198$$

SOLUTION:

$$pK_I = pK_{I=0} - \frac{(2n+1)\,A\,I^{1/2}}{1 + I^{1/2}}$$

$$A = 0.509 \text{ at } 25°C$$

n is defined by $HA^{-n} = H^+ + A^{-(n+1)}$

For pK_1 of H_3PO_4, $n = 0$

$$pK_1 = 2.148 - \frac{(0.509)(0.1)^{\frac{1}{2}}}{1 + (0.1)^{1/2}}$$

$$= 2.148 - 0.122 = 2.026$$

For pK_2 of H_3PO_4, $n = 1$

$$pK_2 = 7.198 - \frac{(3)(0.509)(0.1)^{1/2}}{1 + (0.1)^{1/2}}$$

$$= 7.198 - 0.369 = 6.831$$

6.8 Will 0.01 mol L^{-1} creatine phosphate react with 0.01 mol L^{-1} adenosine disphosphate to produce 0.04 mol L^{-1} creatine and 0.02 mol L^{-1} adenosine triphosphate at 25°C, pH 7, pMg 4? What concentration of ATP can be formed if the other reactants are maintained at the indicated concentrations?

SOLUTION:

$$\text{Creatine P} + H_2O = \text{Creatine} + P \qquad \Delta G^{\circ\prime} = -10.4 \text{ Kcal mol}^{-1}$$

$$ADP + P = ATP + H_2O \qquad \Delta G^{\circ\prime} = 9.5 \text{ kcal mol}^{-1}$$

$$\text{CreatineP} + ADP = \text{Creatine} + ATP \qquad \Delta G^{\circ\prime} = -0.9 \text{ kcal mol}^{-1}$$

$$\Delta G = \Delta G^{\circ} + RT \ln \frac{(\text{Creatine})(ATP)}{(\text{Creatine P})(ADP)}$$

$$= -900 + (1.987)(298) \ln \frac{(0.04)(0.02)}{(0.01)(0.01)}$$

$$= 320 \text{ cal mol}^{-1}$$

Therefore the answer to the first question is no

$$K = e^{-\Delta G^{\circ}/RT} = e^{900/(1.987)(298)}$$

$$= 4.5 = \frac{(\text{Creatine})(ATP)}{(\text{CreatineP})(ADP)}$$

If the reactants are maintained at the indicated concentrations.

$$(ATP) = \frac{4.5\ (\text{Creatine P})(ADP)}{(\text{Creatine})}$$

$$= \frac{4.5\ (0.01)(0.01)}{0.04}$$

$$= 1.1 \times 10^{-2} \text{ mol L}^{-1}$$

6.10 Biochemistry textbooks give $\Delta G^{\circ\prime} = 4.80$ Kcal mol^{-1} for the hydrolysis of ethyl acetate at pH 7 and 25° C. Experiments in acid solution show that

$$\frac{(CH_3\ CH_2\ OH)\ (CH_3\ CO_2\ H)}{(CH_3\ CO_2\ CH_2\ CH_3)} = 14$$

where concentrations are in moles per liter. What is the value of $\Delta G^{\circ\prime}$ obtained from this equilibrium quotient? The pK of acetic acid = 4.60 at 25° C.

SOLUTION:

$$K' = \frac{(CH_3\,CH_2\,OH)[(CH_3\,CO_2\,H)+(CH_3\,CO_2^-)]}{(CH_3\,CO_2\,CH_2\,CH_3)}$$

$$= \frac{(CH_3\,CH_2\,OH)(CH_3\,CO_2\,H)}{(CH_3\,CO_2\,CH_2\,CH_3)}\left[1+\frac{(CH_3\,CO_2^-)}{(CH_3\,CO_2\,H)}\right]$$

$$= 14 \text{ mol L}^{-1}\left[1+\frac{K_{CH_3CO_2H}}{(H^+)}\right]$$

At pH 7

$$K' = 14\left[1+\frac{10^{-4.6}}{10^{-7}}\right] = 3530$$

$$\Delta G^{\circ\prime} = -RT \ln K$$

$$= -(1.987 \text{ cal K}^{-1} \text{ mol}^{-1})(298.15 \text{ K}) \ln 3540$$

$$= -4.84 \text{ Kcal mol}^{-1}$$

6.11 The cleavage of fructose 1,6-diphosphate (FDP) to dihydroxyacetone phosphate (DHP) and glyceraldehyde 3-phosphate (GAP) is one of a series of reactions most organisms use to obtain energy. At 37° C and pH 7, $\Delta G^{\circ\prime}$ for the reaction FDP = DHP + GAP is 5.73 Kcal mol^{-1}. What is $\Delta G^{\circ\prime}$ in an erythrocyte in which (FDP) = 3 μmol L^{-1}, (DPH) = 138 μmol L^{-1}, and (GAP) = 18.5

$\mu mol\ L^{-1}$?

SOLUTION:

$$FDP + H_2O = DHP + GAP$$

$$\Delta G^{o\prime} = \Delta G^{o\prime} + RT \ln \frac{(DHP)(GAP)}{(FDP)}$$

$$= 5730 + (1.987)(310) \ln \frac{(138 \times 10^{-6})(18.5 \times 10^{-6})}{(3 \times 10^{-6})}$$

$$= 1380 \text{ cal mol}^{-1}$$

6.12 From the data of Table 6.3 calculate $\Delta G^{o\prime}$ for

$$ATP + H_2O = AMP + PP$$

SOLUTION:

$ATP + H_2O = ADP + P$	$\Delta G^{o\prime} = -9.5$ kcal mol^{-1}
$ADP + H_2O = AMP + P$	$\Delta G^{o\prime} = -8.8$ kcal mol^{-1}
$2P = PP + H_2O$	$\Delta G^{o\prime} = +8.2$ kcal mol^{-1}
$ATP + H_2O = AMP + PP$	$\Delta G^{o\prime} = -10.1$ kcal mol^{-1}

6.13 0.01 mol L^{-1} NaH_2PO_4 dissolved in 0.2 mol L^{-1} $(CH_3CH_2)_4NCl$ is titrated with $(CH_3CH_2)_4NOH$ at 25 °C. The midpoint of the titration curve is at pH 6.80. An identical solution is made 0.05 mol L^{-1} in $MgCl_2$ and the titration is repeated. This time the midpoint is pH 6.37. What is the dissociation constant of $MgHPO_4$?

SOLUTION:

$$K = \frac{(H^+)\left[(HPO_4^{2-}) + (MgHPO_4)\right]}{(H_2PO_4^-)} = \frac{(H^+)(HPO_4^{2-})}{(H_2PO_4^-)}\left[1 + \frac{(MgHPO_4)}{(HPO_4^{2-})}\right]$$

$$= \frac{(H^+)(HPO_4^{2-})}{(H_2PO_4^-)}\left[1 + \frac{(Mg^{2+})}{K_{MgP}}\right]$$

$$10^{-6.37} = 10^{-6.80}\left[1 + \frac{0.05 \text{ mol L}^{-1}}{K_{MgP}}\right]$$

$$K_{MgP} = 2.96 \times 10^{-2} \text{ mol L}^{-1}$$

6.15 Given $\Delta G° = +11.8$ Kcal mol^{-1} for

$$ATP^{4-} + H_2O = AMP^{2-} + P_2O_7^{4-} + 2H^+$$

Calculate $\Delta G°'$ at pH 7 and 25°C and 0.2 ionic strength. See Table 6.4.

SOLUTION:

$$ATP^{4-} + H_2O \rightleftharpoons AMP^{2-} + P_2O_7^{4-} + 2H^+$$

$$ATP^{4-} \underset{10^{-6.95}}{\rightleftharpoons} HATP^{3-} \qquad AMP^{2-} \underset{10^{-6.45}}{\rightleftharpoons} HAMP^{1-} \qquad P_2O_7^{4-} \underset{10^{-8.95}}{\rightleftharpoons} HP_2O_7^{3-} \underset{10^{-6.12}}{\rightleftharpoons} H_2P_2O_7^{2-}$$

$$K' = \frac{[(AMP^{2-}) + (HAMP^{-})][(P_2O_7^{4-}) + (HP_2O_7^{3-}) + (H_2P_2O_7^{2-})]}{[(ATP^{4-}) + (HATP^{3-})]}$$

$$= \frac{(AMP^{2-})(P_2O_7^{4-})}{(ATP^{4-})} \frac{\left[1 + \frac{(HAMP^{-})}{(AMP^{2-})}\right]\left[1 + \frac{(HP_2O_7^{3-})}{(P_2O_7^{4-})} + \frac{(H_2P_2O_7^{2-})}{(P_2O_7^{4-})}\right]}{\left[1 + \frac{(HATP^{3-})}{(ATP^{4-})}\right]}$$

$$= \frac{(AMP^{2-})(P_2O_7^{4-})(H^+)^2}{(ATP^{4-})} \cdot \frac{\left[1 + \frac{(H^+)}{K_{1AMP}}\right]\left[1 + \frac{(H^+)}{K_{1PP}} + \frac{(H^+)^2}{K_{1PP}K_{2PP}}\right]}{(H^+)^2 \left[1 + \frac{(H^+)}{K_{1ATP}}\right]}$$

$$= e^{\frac{-11,800}{(1.987)(298)}} \frac{\left[1+\frac{10^{-7}}{10^{-6.45}}\right]\left[1+\frac{10^{7}}{10^{-8.95}}+\frac{(10^{-7})^2}{10^{-8.96}\ 10^{-6.12}}\right]}{(10^{-7})^2\left[1+\frac{10^{-7}}{10^{-6.95}}\right]}$$

$$= 1.53 \times 10^7$$

$$\Delta G^{\circ\prime} = -RT \ln K' = -(1.987 \text{ cal K}^{-1}\text{mol}^{-1})(298\text{K}) \ln (1.53 \times 10^7)$$

$$= -9800 \text{ cal mol}^{-1}$$

6.16 The hydrolysis of adenosine triphosphate ATP to adenosine diphosphate ADP and inorganic phosphate at pH 8 and 25°C

$$ATP^{-4} + H_2O = ADP^{-3} + HPO_4^{-2} + H^+$$

has a standard enthalpy change of -3 Kcal mol^{-1}. The standard enthalpy changes of acid dissociation of $HATP^{-3}$, $HADP^{-2}$, and $H_2PO_4^{-1}$ are -2, 0, and +2 Kcal mol^{-1}, respectively. Calculate the standard enthalpy change for the reaction.

$$HATP^{-3} + H_2O = HADP^{-2} + H_2PO_4^-$$

SOLUTION:

$ATP^{4-} + H_2O = ADP^{3-} + HPO_4^{2-} + H^+$	$\Delta H^\circ = -3$ Kcal mol^{-1}
$HATP^{3-} = H^+ + ATP^{4-}$	-2
$H^+ + ADP^{3-} = HADP^{2-}$	0
$H^+ + HPO_4^{2-} = H_2PO_4^{-1}$	- 2
$HATP^{3-} + H_2O = HADP^{2-} + H_2PO_4^{-1}$	$\Delta H^\circ = -7$ Kcal mol^{-1}

6.17 Derive the equations for the number r_A of

of A bound per molecule of P and the number r_B of molecules of B bound per molecule of P for the reactions

$$PA \overset{K_A}{=} P+A$$

$$PB \overset{K_B}{=} P+B$$

where the K's represent dissociation constants. Show that

$$\frac{\partial r_A}{\partial \ln (B)} = \frac{\partial r_B}{\partial \ln (A)}$$

This relation expresses the linkage between the binding of A and B.

SOLUTION:

$$r_A = \frac{(PA)}{(P)+(PA)+(PB)} = \frac{(A)/K_A}{1+(A)/K_A+(B)/K_B}$$

$$r_B = \frac{(PB)}{(P)+(PA)+(PB)} = \frac{(B)/K_B}{1+(A)/K_A+(B)/K_B}$$

$$\frac{\partial r_A}{\partial \ln (B)} = \frac{(B)\,\partial r_A}{\partial (B)} = \frac{-\frac{(B)}{K_B}\,\frac{(A)}{K_A}}{[1+(A)/K_A+(B)/K_B]^2}$$

$$\frac{\partial r_B}{\partial \ln (A)} = \frac{(A)\,\partial r_A}{\partial (A)} = \frac{-\frac{(A)}{K_A}\,\frac{(B)}{K_B}}{[1+(A)/K_A+(B)/K_B]^2}$$

6.18 If n molecules of a ligand L combine with a molecule of protein to form PL_n without intermediate steps, derive the relation between the fractional saturation Y and the concentration of L.

SOLUTION:

$$P + nL = PL_n$$

$$K = \frac{(P)(L)^n}{(PL_n)} \qquad (P)_o = (P) + (PL_n)$$

$$\frac{[(P)_o - (PL_n)](L)^n}{(PL_n)} = K$$

$$(P)_o(L)^n = (PL_n)[K + (L)^n]$$

$$Y = \frac{(PL_n)}{(P)_o} = \frac{1}{1 + K/(L)^n} = \frac{(L)^n/K}{1 + (L)^n/K}$$

This equilibrium represents a cooperative effect in that as soon as one ligand molecule is bound, the other (n−1) ligand molecules are also bound.

6.19 The percent saturation of a sample of human hemoglobin was measured at a series of oxygen partial pressures at 20°C, pH 7.1, 0.3 mol L^{-1} phosphate buffer and 3×10^{-4} mol L^{-1} heme.

P_{O_2} /torr	Percent Saturation
2.95	4.8
5.9	20
8.9	45
18.8	78
22.4	90

Calculate the values of n and K in the Hill equation.

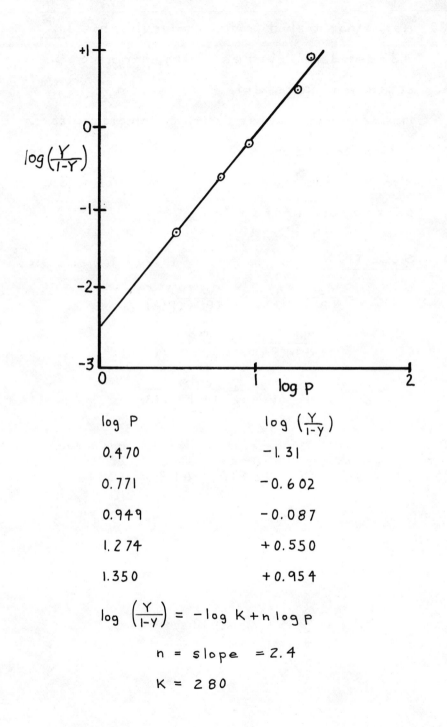

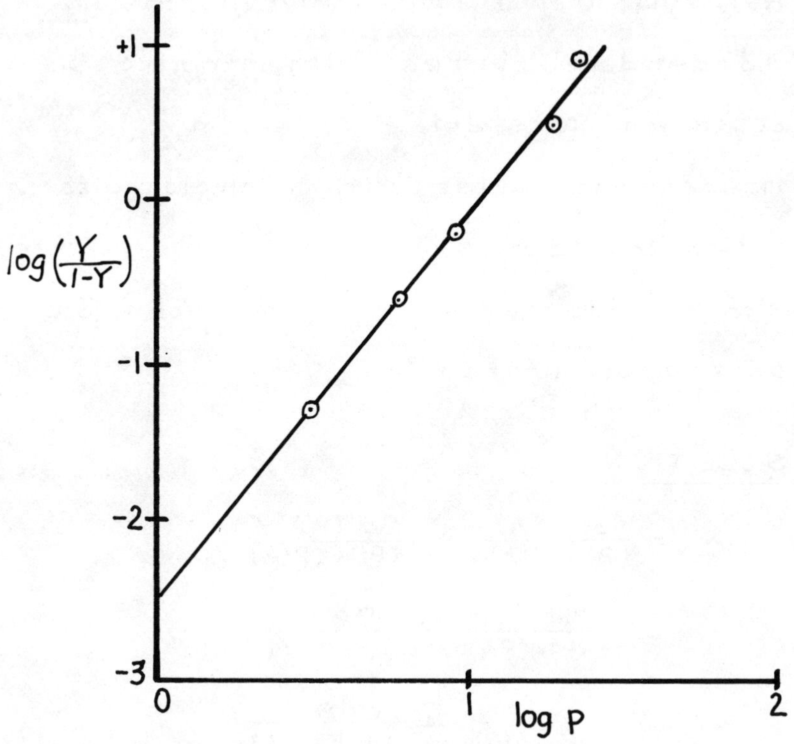

log P	$\log\left(\frac{Y}{1-Y}\right)$
0.470	−1.31
0.771	−0.602
0.949	−0.087
1.274	+0.550
1.350	+0.954

$$\log\left(\frac{Y}{1-Y}\right) = -\log K + n \log P$$

$$n = \text{slope} = 2.4$$

$$K = 280$$

6.20 Assuming that a protein molecule has n_α independent sites with intrinsic dissociation constant K_α and n_β independent sites with intrinsic dissociation constant K_β, what is the expression for the number r of moles of A bound pe mole of protein?

SOLUTION:

$$r = \frac{n_\alpha (PA)}{(P)+(PA)} + \frac{n_\beta (P'A)}{(P')+(P'A)}$$

$$= \frac{n_\alpha}{1+(P)/(PA)} + \frac{n_\beta}{1+(P')/(P'A)}$$

$$= \frac{n_\alpha}{1+K_\alpha/(A)} + \frac{n_\beta}{1+K_\beta/(A)}$$

6.21

	acetic	lactic	bromoacetic
(a) approx. α	0.0424	0.118	0.374
(b) exact α	0.0416	0.112	0.311

6.22

2.3   3.9   6.9   10.0
pK₁   pK₂   pK₃

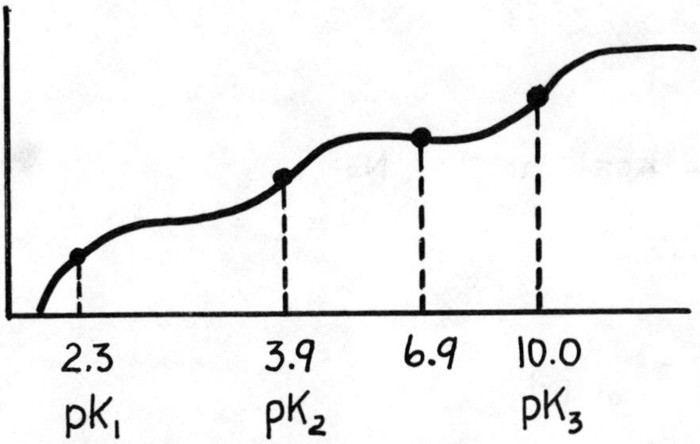

The pH of a solution of monosodium aspartate
is $(pK_2 + pK_3)/2 = 6.9$. The pH of a solution
of disodium aspartate is greater than 10.

Monosodium aspartate
$$\begin{array}{c} CH_2-CO_2^- \ Na^+ \\ | \\ {}^+H_3N-C-CO_2^- \end{array}$$

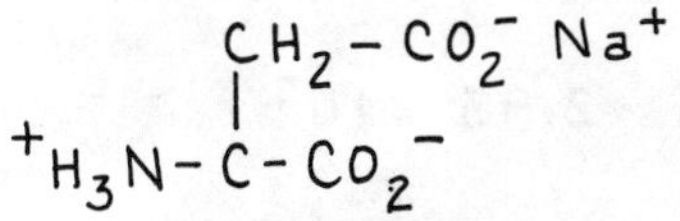

Disodium aspartate
$$\begin{array}{c} CH_2-CO_2^- \ Na^+ \\ | \\ H_2N-C-CO_2^- \ Na^+ \end{array}$$

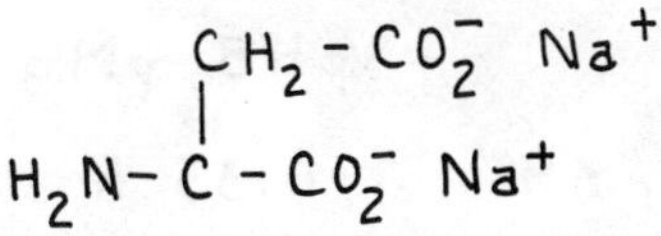

6.23 (a) 7.14   (b) 0.03   (c) 4

6.24 $10^{-4.74} = 1.82 \times 10^{-5}$

6.26 For acetic acid dissociation $\Delta S° = -21.7$ cal
K⁻¹ mol⁻¹. The increase in order is due
to the hydration of the ions that are
formed. For aniline $\Delta S° = -4.4$ cal K⁻¹ mol⁻¹.
This number is smaller because there is
no change in the number of ions.

6.27 4.71

6.28 7.4

6.29 0.063

6.30 $\Delta G = 2.2$ kcal mol^{-1} No

6.31 133 g

6.32 3.88×10^{-2} mol L^{-1}

6.33 4.35

6.34 -7.25 kcal mol^{-1}

6.35 2.93×10^{7}

6.36

	pH 8, pMg 4	pH 8, pMg 8
$\Delta H^{\circ\prime}$	-6.35 kcal mol^{-1}	-4.7 kcal mol^{-1}
$\Delta G^{\circ\prime}$	-10.2 kcal mol^{-1}	-10.6 kcal mol^{-1}

6.37 (a) $K' = K_1 \, [1 + K_{CH}/(H^+)]/[1 + K_{AH}/(H^+)]$

(b) $K_2 \, K_{CH} / K_1 \, K_{AH} = 1$

6.38 25.6%

6.39 $K_1 = k/5$, $K_2 = k/2$, $K_3 = k$, $K_4 = 2k$, $K_5 = 5k$

1/32, 5/32, 10/32, 10/32, 5/32, 1/32

6.40 $$Y = \frac{2 + K_2/P_{O_2}}{2[1 + K_2/P_{O_2} + K_1 K_2/P_{O_2}^2]}$$

CHAPTER 7 Surface Thermodynamics

7.1 Estimate the surface tension of a metal given that the heat of sublimation is 10^5 cal mol^{-1} and that there are 10^{15} atoms cm^{-2} in the surface. It may be assumed that the structure is close packed so that each atom has 12 nearest neighbors. With this information only an approximate answer can be calculated.

SOLUTION:

Sublimation transfers atoms from the bulk solid phase to the gas phase. To set free an atom surrounded by 12 neighbors it is necessary to break 12 bonds, but since each bond connects two neighboring atoms, to set free N atoms 12N/2 bonds have to be broken. On the average new surface requires the breaking of 3 bonds per atom. Thus the energy per m^2 of surface is

$$\frac{(3)\left(\frac{10^5}{6}\ \text{cal mol}^{-1}\right)(4.184\ \text{J cal}^{-1})(10^{15}\ \text{cm}^{-2})(10^4\ \text{cm}^2\ \text{m}^{-2})}{6.02 \times 10^{23}\ \text{mol}^{-1}}$$

$$= 3.5\ \text{J m}^{-2}$$

7.2 The surface tension of toluene at 20° C is 0.0284 N m^{-1}, and its density at this temperature is 0.866 g cm^{-3}. What is the radius of the largest capillary that will permit the liquid to rise 2 cm?

SOLUTION:

$$\gamma = \frac{1}{2} h\rho g r$$

$$r = \frac{2\gamma}{\rho g h} = \frac{2(0.284\ \text{N m}^{-1})}{(0.866 \times 10^3\ \text{kg m}^{-3})(9.80\ \text{m s}^{-2})(0.02\ \text{m})}$$

$$= 3.35 \times 10^{-4}\ \text{m}$$

$$= 3.35 \times 10^{-2}\ \text{cm}$$

7.3 Calculate the vapor pressure of a water droplet at 25 °C that has a radius of 2 nm. The vapor pressure of a flat surface of water is 23.76 torr at 25 °C.

SOLUTION:

$$\ln\left(\frac{P}{P^\circ}\right) = \frac{2V_m\gamma}{rRT}$$

$$\ln\left(\frac{P}{23.76\ \text{torr}}\right) = \frac{2(18.016 \times 10^{-3}\ \text{kg mol}^{-1})(0.07197\ \text{N m}^{-1})}{(2 \times 10^{-9}\ \text{m})(8.314\ \text{J K}^{-1}\ \text{mol}^{-1})(298.15\ \text{K})} \times \frac{1}{(10^3\ \text{kg m}^{-3})}$$

$$P = 40.1\ \text{torr}$$

7.4 The surface tensions of 0.05 mol kg^{-1} and 0.127 mol kg^{-1} solutions of phenol in water are 67.7 and 60.1 mN m^{-1}, respectively, at 20 °C. What is the surface concentration $\Gamma_2^{(1)}$ in the range 0-0.05 mol kg^{-1} and 0.05 - 0.127 mol kg^{-1}, assuming that the phenol concentrations can be treated as activities. The surface tension of water at 20 °C is 72.7 mN m^{-1}.

SOLUTION:

$$\Gamma_2^{(1)} = \frac{-a_2}{RT} \frac{\partial \gamma}{\partial a_2}$$

In the range 0–0.05 mol kg^{-1}

$$\frac{\partial \gamma}{\partial a_2} = \frac{67.7-72.7}{0.05} = -100 \times 10^{-3} \text{ N m}^{-1}$$ (since the activity a_2 is taken as dimensionless)

$$\Gamma_2^{(1)} = -\frac{(0.025)(-100 \times 10^{-3} \text{ N m}^{-1})}{(8.314 \text{ J K}^{-1} \text{ mol}^{-1})(293 \text{ K})}$$

$$= 1.0 \times 10^{-6} \text{ mol m}^{-2}$$

In the range 0.05–0.127 mol kg^{-1}

$$\frac{\partial \gamma}{\partial a_2} = \frac{60.1-67.7}{0.077} = -99 \times 10^{-3} \text{ N m}^{-1}$$

$$\Gamma_2^{(1)} = -\frac{[(0.05+0.127)/2](-99 \times 10^{-3} \text{ N m}^{-1})}{(8.314 \text{ J K}^{-1} \text{mol}^{-1})(293 \text{ K})}$$

$$= 3.6 \times 10^{-6} \text{ mol m}^{-2}$$

7.5 (a) The surface tension of water against air at 1 atm is given in the following table for various temperatures:

$t/°C$	20	22	25	28	30
$\gamma/\text{N m}^{-1}$	0.07275	0.07244	0.07197	0.07150	0.07118

Calculate the surface enthalpy at 25°C in calories per square centimeter. (b) If a finely divided solid whose surface is covered with a very thin layer of water is dropped into a container of water at the same temperature, heat will be evolved. Calculate the heat evolution for 10 grams of a powder having a surface area of 200 $m^2\ g^{-1}$ [W. D. Harkins and G. Jura, J. Am. Chem. Soc., 66, 1362 (1944)].

SOLUTION:

(a) $$h^\sigma = \gamma - T\left(\frac{\partial \gamma}{\partial T}\right)_P$$

$$= 0.07197\ J\ m^{-2} - (298.15\ K)(-1.48 \times 10^{-4}\ J\ m^{-2} K^{-1})$$

$$= 0.1161\ J\ m^{-2}$$

$$= \frac{(0.1161\ J\ m^{-2})(10^{-4}\ m^2\ cm^{-2})}{4.184\ J\ cal^{-1}}$$

$$= 2.78 \times 10^{-6}\ cal\ cm^{-2}$$

(b) $$(10 g)(200\ m^2 g^{-1})(10^4\ cm^2\ m^{-2})(2.78 \times 10^{-6}\ cal\ cm^{-2})$$

$$= 55.6\ cal$$

7.7 A protein with a molar mass of 60,000 $g\ mol^{-1}$ forms a perfect gaseous film on water. What area of film per milligram of protein will produce a pressure of 0.005 $N m^{-1}$ at 25°C?

SOLUTION:

$$\sigma = \frac{RT}{\Pi}$$

$$\text{area} = \frac{g N_A \sigma}{M} = \frac{g N_A RT}{M \Pi}$$

$$= \frac{(10^{-6} Kg)(6.022 \times 10^{23} mol^{-1})(1.38 \times 10^{-23} J K^{-1})(298 K)}{(60\ Kg\ mol^{-1})(0.005\ N\ m^{-1})}$$

$$= 82.5 \times 10^{-4}\ m^2$$

$$= 82.5\ cm^2$$

7.8 The acid $CH_3(CH_2)_{13}CO_2H$ forms a nearly perfect gaseous monolayer on water at 25°C. Calculate the weight of acid per 100 cm^2 required to produce a film pressure of 10^{-3} N m^{-1}.

SOLUTION:

$$\pi \mathcal{A} = \frac{N}{N_A} RT = \frac{g\,RT}{M}$$

$$g = \frac{\pi \mathcal{A} M}{RT}$$

$$= \frac{(10^{-3}\,N\,m^{-1})(10^{-2}\,m^2)(244 \times 10^{-3}\,kg\,mol^{-1})}{(8.314\,J\,K^{-1}\,mol^{-1})(298.15\,K)}$$

$$= 9.8 \times 10^{-10}\ kg = 9.8 \times 10^{-7}\ g$$

7.9 A monomolecular film obeys the following equation of state

$$\pi(\sigma - \mathcal{B}) = kT$$

where $\mathcal{B}$ may be interpreted as the effective area occupied by a single molecule. This equation may also be written

$$\pi = \frac{kT}{B}\,\frac{\theta}{1-\theta}$$

where θ is the fraction of the surface occupied. Derive the relation between pressure in the vapor phase and θ.

SOLUTION:

$$d\Pi = RT\,\Gamma_2^{(1)}\,d\ln P$$

$$\frac{d\Pi}{d\ln P} = RT\,\Gamma_2^{(1)} = \frac{RT}{N_A\sigma} = \frac{kT}{\sigma} = \frac{kT\theta}{\beta}$$

The last form is obtained using $\sigma = \beta/\theta$

$$d\ln P = \frac{\beta}{kT\theta}\,d\Pi$$

$$\Pi = \frac{kT}{\beta}\left[\theta(1-\theta)^{-1}\right]$$

$$d\Pi = \frac{kT}{\beta}\left[\frac{\theta}{(1-\theta)^2} + \frac{1}{1-\theta}\right]d\theta = \frac{kT}{\beta}\left[\frac{d\theta}{(1-\theta)^2}\right]$$

$$\int d\ln P = \int\frac{d\theta}{\theta(1-\theta)^2} = \frac{1}{(1-\theta)} - \ln\frac{1-\theta}{\theta}$$

$$\ln P = \ln\frac{\theta}{1-\theta} + \frac{1}{1-\theta} + \text{const}$$

$$P = K\frac{\theta}{1-\theta}\,e^{1/(1-\theta)}$$

7.11 Two gases, A and B, compete for the binding sites on the surface of an adsorbent. Show that the fraction of the surface covered by A molecules is

$$\theta_A = \frac{b_A P_A}{1 + b_A P_A + b_B P_B}$$

where b_A and b_B are constants.

SOLUTION

For A $\theta_A r_A = k_A (1-\theta_A - \theta_B) P_A$ (1)

$\theta_B r_B = k_B (1-\theta_A - \theta_B) P_B$ (2)

$$\frac{\theta_A r_A}{k_A P_A} = (1-\theta_A - \theta_B) = \frac{\theta_B r_B}{k_B P_B} \quad (3)$$

Using equation 3 to eliminate θ_B from equation 1,

$$\theta_A r_A = k_A \left(1-\theta_A - \frac{r_A k_B P_B}{r_B k_A P_A}\theta_A\right) P_A$$

Solving for θ_A

$$\theta_A = \frac{\frac{k_A}{r_A} P_A}{1+\frac{k_A}{r_A} P_A + \frac{k_B}{r_B} P_B}$$

7.12 The following table gives the number of milliliters (v) of nitrogen (reduced to 0° C and 1 atm) adsorbed per gram of active carbon at 0° C at a series of pressures:

P/torr	3.93	12.98	22.94	34.01	56.23
$v/cm^3 g^{-1}$	0.987	3.04	5.08	7.04	10.31

Plot the data according to the Langmuir isotherm, and determine the constants.

SOLUTION:

$$\frac{1}{v} = \frac{1}{v_m} + \frac{k'}{v_m P}$$

$$\text{Intercept} = 0.025 \text{ g cm}^{-3}$$

$$V_m = 1/(0.025 \text{ g cm}^{-3})$$

$$= 40 \text{ cm}^3 \text{ g}^{-1}$$

$$\text{Slope} = \frac{(1.00 - 0.025) \text{ g cm}^{-3}}{0.25 \text{ torr}^{-1}}$$

$$= 3.9 \text{ g cm}^{-3} \text{ torr}$$

$$= \frac{R'}{V_m} = \frac{R'}{40 \text{ cm}^3 \text{ g}^{-1}}$$

$$R = 156 \text{ torr}$$

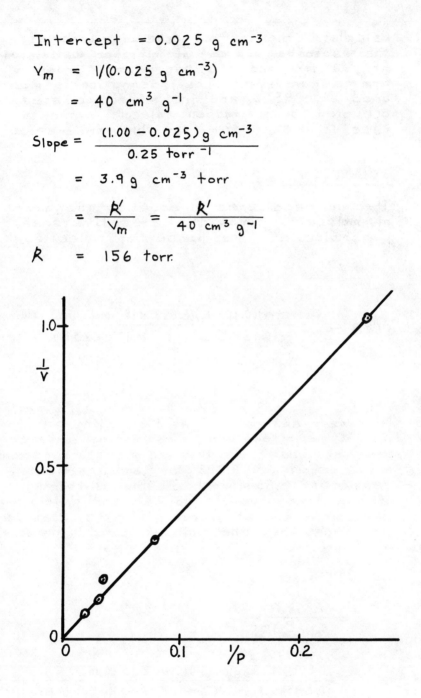

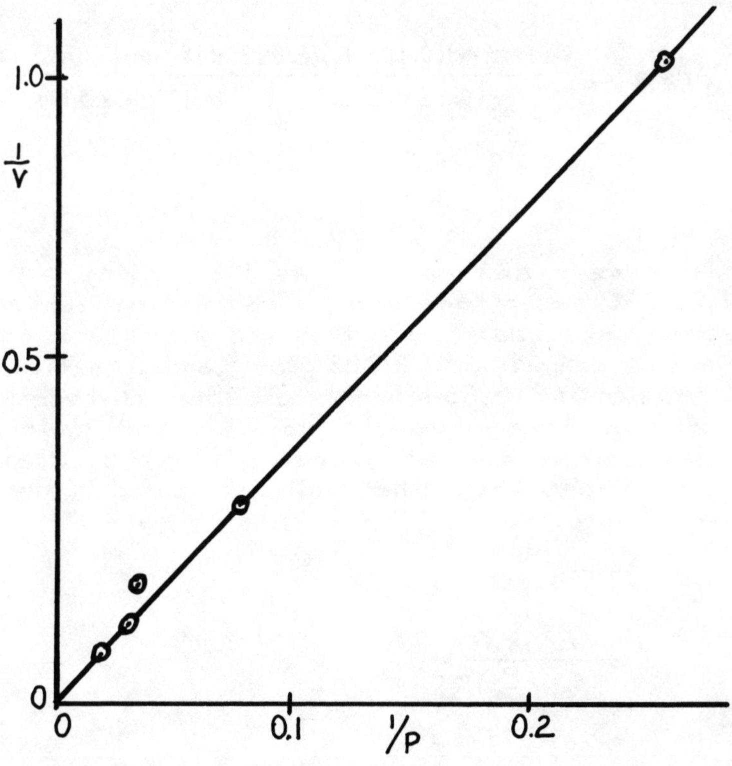

7.13 Calculate the surface area of a catalyst that absorbs 103 cm^3 of nitrogen (calculated at 760 torr and 0°C) per gram in order to form a monolayer. The adsorption is measured at −195°C, and the effective area occupied by a nitrogen molecule on the surface is $16.2 \times 10^{-20}\ m^2$ at this temperature.

SOLUTION:

The surface area is equal to the number of molecules times the effective area per molecule. The number of molecules is

$$\frac{PV N_A}{RT}$$

$$\text{Area} = \frac{(1\ \text{atm})(0.103\ \text{L})(6.022 \times 10^{23}\ \text{mol}^{-1})(16.2 \times 10^{-20}\ \text{m}^2)}{(0.082\ \text{L atm K}^{-1}\ \text{mol}^{-1})(273\ \text{K})}$$

$$= 449\ \text{m}^2$$

7.15 Nitrogen adsorption, at its boiling point −195.8°C, is often used in estimating surface areas of solids. At this temperature nitrogen has a density of 0.808 g cm^{-3} and a surface tension of 8.85 mN m^{-1}. If the isotherm is of the type shown in Fig 7.8b and hysteresis is encountered at about $P/P° = 0.5$, what does this imply about the radii of pores in the solid?

SOLUTION:

$$r = \frac{2 V_m \gamma}{(-\ln \frac{P}{P°}) RT}$$

$$= \frac{2\,(14 \times 10^{-3}\ \text{kg mol}^{-1})(0.00885\ \text{N m}^{-1})}{(0.693)(8.314\ \text{J K}^{-1}\ \text{mol}^{-1})(77.4\ \text{K})(0.808 \times 10^{3}\ \text{kg m}^{-3})}$$

$= 0.69 \times 10^{-9}$ m

7.16 0.0233 N m^{-1}

7.17 (a) 15.72 cm (b) 0.786 cm

7.18 (a) 15.2×10^{-10} m (b) 97

7.19 99.2×10^{-10} m = 9.92 nm

7.21 (a) A cooling effect because energy is required to move molecules to the surface.

(b) 0.1187 J m^{-2}

7.22 0.0269 J m^{-2}

7.23 244 m

7.24 4.04×10^{-6} mol m^{-2}

7.25 $\Pi = \frac{kT}{\sigma_0} \ln\left(\frac{1}{1-\theta}\right)$

7.26 (a) $v_m = 38.2$ mm^3 $k' = 6.11$ torr

(b) 30.4 mm^3

7.27 39.6 L

7.28 20.3 L at 25°C and 1 atm

7.29 8650 cal mol^{-1}

7.30 7.06 kcal mol^{-1}

7.31 1.51 nm

PART TWO: QUANTUM CHEMISTRY

CHAPTER 8 Quantum Theory

8.1 A hollow box with an opening of 1 cm^2 area is heated electrically. (a) What is the total energy emitted per second at 800 K? (b) How much energy is emitted per second if the temperature is 1600 K? (c) How long would it take the radiant energy emitted at this temperature, 1600 K, to melt 1000 g of ice?

<u>SOLUTION:</u>

(a) $I = \sigma T^4 = (5.67 \times 10^{-8}\ J s^{-1} m^{-2} K^{-4})(800\ K)^4$

$$= 2.32 \times 10^4\ J m^{-2} s^{-1}$$

(b) $I = \sigma T^4 = (5.67 \times 10^{-8}\ J s^{-1} m^{-2} K^{-4})(1600\ K)^4$

$$= 37.2 \times 10^4\ J m^{-2} s^{-1}$$

(c) $\Delta H = (79.7\ cal\ g^{-1})(1000\ g\ Kg^{-1})(4.184\ J\ cal^{-1})$

$$= 3.33 \times 10^5\ J$$

$$t = \frac{3.33 \times 10^5\ J}{37.2\ J s^{-1}}$$

$$= 8950\ s$$

8.2 Calculate the ratio of the intensities of light of 500 nm wavelength from cavities of 1000 K and 5000 K.

SOLUTION:

$$\frac{M_{\lambda 1000K}}{M_{\lambda 5000K}} = \frac{e^{\frac{1.43879 \times 10^{-2}\,\text{mK}}{(500 \times 10^{-9}\,\text{m})(5000\text{K})}} - 1}{e^{\frac{1.43879 \times 10^{-2}\,\text{mK}}{(500 \times 10^{-9}\,\text{m})(1000\text{K})}} - 1}$$

$$= 1.00 \times 10^{-10}$$

8.3 Calculate the wavelengths (in micrometers) of the first three lines of the Paschen series for atomic hydrogen.

SOLUTION:

$$\lambda = \frac{1}{R} \frac{n_1^2 n_2^2}{(n_2^2 - n_1^2)} = \frac{1}{R} \frac{9 n_2^2}{(n_2^2 - 9)}$$

For $n_2 = 4$ $\quad \lambda = \frac{(9)(16)}{(109677.58\ \text{cm}^{-1})(7)} = 1.8756\ \mu\text{m}$

For $n_2 = 5$ $\quad \lambda = \frac{(9)(25)}{(109677.58\ \text{cm}^{-1})(16)} = 1.2822\ \mu\text{m}$

For $n_2 = 6$ $\quad \lambda = \frac{(9)(36)}{(109677.58\ \text{cm}^{-1})(27)} = 1.0941\ \mu\text{m}$

8.5 Calculate the wavelength of light emitted when an electron falls from the n = 100 orbit to the n = 99 orbit of the hydrogen atom. Such species are know as high Rydberg atoms. They are detected in astronomy and are increasingly studied in the laboratory.

SOLUTION:

$$E_{100} - E_{99} = -1.0967758 \times 10^5 \text{cm}^{-1}\left(\frac{1}{100^2} - \frac{1}{99^2}\right)$$

$$= 22.2690 \text{ cm}^{-1}$$

$$\lambda = 0.0449055 \text{ cm}$$

8.6 What potential difference is required to accelerate a singly charged gas ion in a vacuum so that it has (a) a Kinetic energy equal to that of an average gas molecule at 25°C. (b) an energy equivalent to 20 Kcal mol^{-1}?

SOLUTION:

(a) It has been shown earlier that the Kinetic energy of an average gas molecule is [(3/2) kT].

$$Ee = \frac{3}{2} kT$$

$$E = \frac{3\,(1.381 \times 10^{-23} \text{J K}^{-1})(298 \text{ K})}{2(1.602 \times 10^{-19} \text{ C})}$$

$$= 0.0385 \text{ V}$$

(b) $$Ee = \frac{(20{,}000 \text{ cal mol}^{-1})(4.184 \text{ J cal}^{-1})}{6.022 \times 10^{23} \text{ mol}^{-1}}$$

$$= 1.3896 \times 10^{-19} \text{ J}$$

$$E = \frac{1.3896 \times 10^{-19}\ J}{1.602 \times 10^{-19}\ C}$$

$$= 0.867\ V$$

Since this type of conversion is done so often it is worth remembering the conversion factor 23.060 Kcal mol^{-1} eV^{-1}.

$$\frac{20\ Kcal\ mol^{-1}}{23.060\ Kcal\ mol^{-1}\ eV^{-1}} = 0.867\ eV$$

8.7 Calculate the velocity of an electron that has been accelerated by a potential difference of 1000 V.

SOLUTION:

$$Ee = \frac{1}{2} m v^2$$

$$v = \left(\frac{2Ee}{m}\right)^{1/2}$$

$$= \left[\frac{2(1000\ V)(1.602 \times 10^{-19}\ C)}{9.110 \times 10^{-31}\ Kg}\right]^{1/2}$$

$$= 1.875 \times 10^{7}\ ms^{-1}$$

As higher accelerating potentials are used it becomes necessary to use the relativistic mass of the electron as discussed in the footnote in Section 1.8.

8.8 Electrons are accelerated by a 1000 V potential drop. (a) Calculate the de Broglie wavelength. (b) Calculate the wavelength of the X-rays that could be produced when those electrons strike a solid.

SOLUTION:

(a) $$E e = \frac{1}{2} m v^2$$

$$v = \left(\frac{2 E e}{m}\right)^{1/2}$$

$$= \left[\frac{2 (1000 V)(1.602 \times 10^{-19} c)}{9.110 \times 10^{-31} Kg}\right]^{1/2}$$

$$= 1.875 \times 10^7 m s^{-1}$$

$$\lambda = \frac{h}{mv}$$

$$= \frac{6.626 \times 10^{-34} Js}{(9.110 \times 10^{-31} Kg)(1.875 \times 10^7 m s^{-1})}$$

$$= 0.0387 nm$$

(b) $$E e = hc/\lambda$$

$$\lambda = \frac{hc}{E e}$$

$$= \frac{(6.626 \times 10^{-34} Js)(2.998 \times 10^8 m s^{-1})}{(1000 V)(1.602 \times 10^{-19} c)}$$

$$= 1.24 nm$$

8.10 (a) Calculate the energy levels for $n=1$ and $n=2$ for an electron in a potential well of width 0.5 nm with infinite barriers on either side. The energies should be expressed in Kcal mol^{-1} (b) If an electron makes a transition from $n=2$ to $n=1$ what will be the wavelength of the radiation emitted?

SOLUTION:

(a) $$E = \frac{n^2 h^2}{8 m a^2}$$

$$= \frac{(6.626 \times 10^{-34} \text{ Js})^2}{8(9.110 \times 10^{-31} \text{ Kg})(5 \times 10^{-10} \text{ m})^2}$$

$$= 2.4096 \times 10^{-19} \text{ J}$$

$$= \frac{(2.4096 \times 10^{-19} \text{J})(6.022 \times 10^{23} \text{ mol}^{-1})}{(4.184 \text{ J cal}^{-1})(1000 \text{ cal Kcal}^{-1})}$$

$$= 3.47 \text{ Kcal mol}^{-1} \quad \text{for } n = 1$$

For $n = 2$

$$E = 4(34.7 \text{ Kcal mol}^{-1})$$

$$= 138.8 \text{ Kcal mol}^{-1}$$

(b) $$\Delta E = (2.4096 \times 10^{-19} \text{ J})(4-1)$$

$$\Delta E = \frac{hc}{\lambda}$$

$$(2.4096 \times 10^{-19} \text{ J})(4.1) = \frac{(6.626 \times 10^{-34} \text{ Js})(2.998 \times 10^{8} \text{ ms}^{-1})}{\lambda}$$

$$\lambda = 274.8 \text{ nm}$$

8.11 Calculate the degeneracies of the first three levels for a particle in a cubical box.

SOLUTION:

$$E = \frac{h^2}{8ma^2}(n_x^2 + n_y^2 + n_z^2) \quad n_x, n_y\ n_z \text{ are } 1, 2, 3\ldots$$

n_x	n_y	n_z	$E/\frac{h^2}{8ma^2}$	
1	1	1	3	Degeneracy = 1
2	1	1		
1	2	1	6	Degeneracy = 3
1	1	2		
2	2	1		
2	1	2	9	Degeneracy = 3
1	2	2		

8.12 Show that the function $\psi = 8e^{5x}$ is an eigenfunction of the operator d/dx. What is the eigenvalue?

SOLUTION:

$$\frac{d\psi}{dx} = \frac{d}{dx} 8e^{5x}$$

$$= 40e^{5x} = 5\psi$$

Thus the eigen value is 5.

8.14 How many Kcal mol^{-1} are required to excite a hydrogen atom from n = 1 to n = 2? How many times larger than the translational energy of a hydrogen atom at room temperature is this?

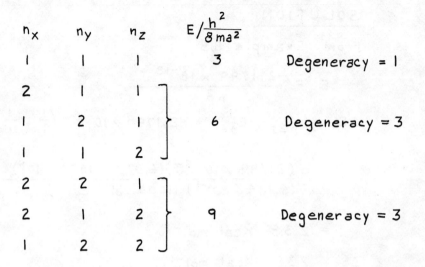

SOLUTION:

From Example 8.5

$$E = \frac{-2.1799 \times 10^{-18}\ \text{J}}{n^2}$$

$$\Delta E = E_{n=2} - E_{n=1} = -2.1799 \times 10^{-18}\ \text{J} \left(\frac{1}{4} - \frac{1}{1}\right)$$

$$= \frac{3(2.1799 \times 10^{-18}\ \text{J})(6.0220 \times 10^{23}\ \text{mol}^{-1})}{4(4.184\ \text{J cal}^{-1})(1000\ \text{cal Kcal}^{-1})}$$

$$= 235\ \text{Kcal mol}^{-1}$$

$$\frac{\Delta E}{\frac{3}{2}RT} = \frac{235\ \text{Kcal mol}^{-1}}{\frac{3}{2}(1.987 \times 10^{-3}\ \text{Kcal mol}^{-1})(298\ \text{K})}$$

$$= 265$$

8.15 What are the wavelengths of the first line in the Balmer series for H (atomic mass 1.007825 g mol^{-1}) and D (atomic mass 2.01410 g mol^{-1})?

SOLUTION:

$$\frac{1}{\lambda} = RH\left(\frac{1}{2^2} - \frac{1}{3^2}\right)$$

$$\lambda = \frac{1}{(1.0967758 \times 10^{7}\ \text{m}^{-1})(1/4 - 1/9)}$$

$$= 656.4696\ \text{nm}$$

$$m_H = \frac{1.007825 \times 10^{-3}\ \text{kg mol}^{-1}}{6.022045 \times 10^{23}\ \text{mol}^{-1}} = 1.673559 \times 10^{-27}\ \text{kg}$$

$$m_e = 0.000\,911 \times 10^{-27}\ \text{kg}$$

$$m_p = 1.672648 \times 10^{-27}\ \text{kg}$$

$$m_D = \frac{2.01410 \times 10^{-3}\ \text{kg mol}^{-1}}{6.022045 \times 10^{23}\ \text{mol}^{-1}} = 3.344545 \times 10^{-27}\ \text{kg}$$

$$m_e = 0.000911 \times 10^{-27}\ \text{kg}$$

$$m_d = 3.34363 \times 10^{-27}\ \text{kg}$$

$$R_H = \text{const.}\ \frac{m_e\, m_p}{m_e + m_p}$$

$$R_D = \text{const.}\ \frac{m_e\, m_d}{m_e + m_d}$$

$$R_D = R_H\ \frac{m_d\,(m_e + m_p)}{m_p\,(m_e + m_d)}$$

$$= (1.0967758 \times 10^{7}\ \text{m}^{-1})\frac{(3.34363)(1.673559)}{(1.672648)(3.344545)}$$

$$= 1.097073 \times 10^{7}\ \text{m}^{-1}$$

$$\lambda_D = \frac{1}{(1.097073 \times 10^{7})(1/4 - 1/9)}$$

$$= 656.292\ \text{nm}$$

8.16 Calculate the Bohr radius with the reduced mass of the hydrogen atom.

SOLUTION:

$$\mu = \frac{1}{\frac{1}{m_e} + \frac{1}{m_p}}$$

$$= \frac{1}{\frac{1}{9.109534 \times 10^{-31}\ Kg} + \frac{1}{1.6726485 \times 10^{-27} Kg}}$$

$$= 9.104575 \times 10^{-31}\ Kg$$

$$a_0 = \frac{h^2\ (4\pi\epsilon_0)}{4\pi^2\ m_e\ e^2} = \frac{h^2\ \epsilon_0}{\pi\ m_e\ e^2}$$

$$= \frac{(6.626176 \times 10^{-34} Js)^2 (8.85418782 \times 10^{-12} C^2 N^{-1} m^{-2})}{\pi (9.104575 \times 10^{-31}\ Kg)(1.6021892 \times 10^{-19}\ C)^2}$$

$$= 0.0529465\ nm$$

8.17 Calculate the average distance between the electron and nucleus of a hydrogen-like atom in the 1s state.

SOLUTION:

$$\langle r \rangle = \int_0^\infty \psi^* r\ \psi\ 4\pi\ r^2\ dr$$

$$\psi_{1s} = \frac{1}{\sqrt{\pi}} \left(\frac{Z}{a_0}\right)^{3/2} e^{-Zr/a_0}$$

$$\langle r \rangle = \frac{4Z^3}{a_0^3} \int_0^\infty e^{-2Zr/a_0}\ r^3\ dr$$

$$= \frac{3}{2}\ \frac{a_0}{Z}$$

8.18 Show that for a 1s orbital of a hydrogen-like atom the most probable distance from proton to electron is a_0/Z.

SOLUTION:

$$\psi_{1s} \propto e^{-Zr/a_0}$$

Probability density of $r \propto r^2 \psi_{1s}^2 = r^2 \times e^{-2Zr/a_0}$. Setting the derivative of the probability density equal to zero,

$$r^2 e^{-2Zr/a_0} \left(\frac{-2Z}{a_0}\right) + 2re^{-2Zr/a_0} = 0$$

$$r = \frac{a_0}{Z}$$

8.19 For the wave function

$$\psi = \begin{vmatrix} \psi_A(1) & \psi_A(2) \\ \psi_B(1) & \psi_B(2) \end{vmatrix}$$

show that (a) the interchange of two columns changes the sign of the wave function, (b) the interchange of two rows changes the sign of the wave function, and (c) the two electrons cannot have the same spin orbital.

SOLUTION:

(a) The interchange of two columns yields

$$\begin{vmatrix} \psi_A(2) & \psi_A(1) \\ \psi_B(2) & \psi_B(1) \end{vmatrix} = \psi_A(2)\,\psi_B(1) - \psi_A(1)\,\psi_B(2)$$

This is the wave function for the atom with the two electrons interchanged, and it is the negative of the function given in the problem, as required by the antisymmetrization principle.

(b) The interchange of two rows yields

$$\begin{vmatrix} \psi_B(1) & \psi_B(2) \\ \psi_A(1) & \psi_A(2) \end{vmatrix} = \psi_A(2)\,\psi_B(1) - \psi_A(1)\,\psi_B(2)$$

which is the negative of the wave function given in the problem. Thus the interchange of two columns or two rows is equivalent to the interchange of two electrons between orbitals.

(c) If two electrons are represented by the same spin orbitals, the determinant.

$$\begin{vmatrix} \psi_A(1) & \psi_A(2) \\ \psi_A(1) & \psi_A(2) \end{vmatrix} = 0$$

is in accord with the principle that two electrons in an atom cannot have four identical quantum numbers.

8.22 Calculate the ionization potential for H(g) from the energy given in Example 8.5.

SOLUTION

$$E = 13.60580\,\text{V}\left(\frac{\mu_H}{m_e}\right)$$

$$= 13.60580\,\text{V}\ \frac{9.104576 \times 10^{-31}\ \text{kg}}{9.109534 \times 10^{-31}\ \text{kg}}$$

$$= 13.59840\ \text{V}$$

8.23 The first ionization potential for atomic lithium is 5.39 V ($Li = Li^{+} + e$). The second ionization potential is 75.62 V ($Li^{+} = Li^{+2} + e$). Calculate the wavelengths for the convergence limits indicated by these potentials.

SOLUTION:

$$Ee = hc/\lambda$$

$$\lambda_1 = \frac{hc}{Ee} = \frac{(6.626 \times 10^{-34}\ \text{Js})(2.998 \times 10^{8}\ \text{ms}^{-1})}{(5.39\ \text{V})\ (1.602 \times 10^{-19}\ \text{c})}$$

$$= 230.1\ \text{nm}$$

$$\lambda_2 = \frac{(6.626 \times 10^{-34}\ \text{Js})\ (2.998 \times 10^{8}\ \text{ms}^{-1})}{(75.62\ \text{V})\ (1.602 \times 10^{-19}\ \text{c})}$$

$$= 16.40\ \text{nm}$$

8.24 857.7 watts

8.25 290 nm

8.26 334 cal min^{-1}

8.27 $2.74 \times 10^{14}\ \text{s}^{-1}$ 1094 nm

8.28 4052, 2626 nm

8.29 5.93×10^5 m s^{-1}

8.30 0.130 nm

8.31 0.00549 nm

8.32 0.146 nm

8.33 (a) 14 (b) 1.37×10^8

8.34 $E = n^2 h^2/8 m \ell^2$

8.35 34.7, 139, 312 kcal mol^{-1}

8.36 1, 3, 3, 3

8.37 $-6a$

8.38 $1/k$

8.39 $6a_0$

8.40 $\langle x \rangle = 0$ $\langle p \rangle = 0$

8.41

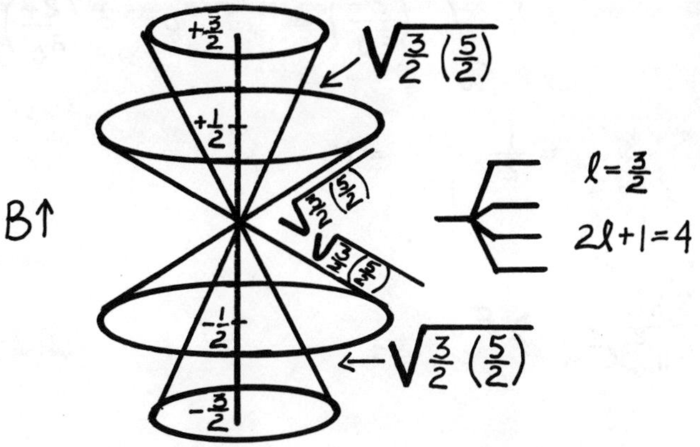

8.42 (a) 243 nm (b) 6.80 V

8.43 109,677.57 cm^{-1}, compared with the experimental value of 109,677.58 cm^{-1}

8.44

$$\langle \psi_{1s} | \psi_{2s} \rangle = (\text{const}) \int_0^\infty r^2 \left(2 - \frac{r}{a_0}\right) e^{-\frac{r}{a_0}} e^{-\frac{r}{2a_0}} dr$$

$$= (\text{const}) \int_0^\infty (2x^2 - x^3) e^{-\frac{3}{2}x} dx \quad \text{where } x = r/a_0$$

$$= (\text{const}) \left(\frac{32}{27} - \frac{32}{27}\right) = 0$$

8.45

$$\langle \psi_{1s} | \psi_{1s} \rangle = \frac{4}{\pi a_0^3} \int_0^\infty r^2 e^{-2r/a_0} dr$$

$$= \frac{1}{2} \int_0^\infty \left(\frac{2r}{a_0}\right)^2 e^{-2r/a_0} \, d\left(\frac{2r}{a_0}\right)$$

$$= \frac{1}{2} \frac{2!}{1^3} = 1$$

8.46 $\frac{5h^2}{4\pi^2 \ell^2 m} \geqslant E_o$

The true value is $h^2/8m\ell^2$

8.47
$$\psi = \frac{1}{\sqrt{6}} \begin{vmatrix} 1s\alpha(1) & 1s\beta(1) & 2s\alpha(1) \\ 1s\alpha(2) & 1s\beta(2) & 2s\alpha(2) \\ 1s\alpha(3) & 1s\beta(3) & 2s\alpha(3) \end{vmatrix}$$

8.48 $\psi = 1s\alpha(1)\ 1s\beta(2) - 1s\alpha(2)\ 1s\beta(1)$

Interchanging coordinates gives

$$1s\alpha(2)\ 1s\beta(1) - 1s\alpha(1)\ 1s\beta(2) = -\psi$$

8.49 The inner electrons (1s) screen the nucleus, but since the probility density of the 2s orbital is greater near the nucleus than that of the 2p, the 2s electrons are not screened as much as the 2p electrons and therefore are bound more tightly (have a lower energy).

8.50 s orbitals (spherically symmetric): H, He, Li, Be, Na, Mg

p orbitals (asymmetric): B, C, N, O, F, Ne, Al, Si, P, S, Cl, Ar

8.51 $E = (13.61\,V)Z^2$

Ion	Z	E/V
He^+	2	54.44
Li^{2+}	3	122.49
Be^{3+}	4	217.76
B^{4+}	5	340.25
C^{5+}	6	489.96

8.52 54.4 V 164 nm 121 nm

8.53 91.58 nm

8.54 13.6627, 13.59844 eV

CHAPTER 9 Symmetry

9.1

E, C_2 (z), σ_v (xz), $\sigma_v{}'$ (yz)

9.2

E, C_3, $3\sigma_v$

9.3

E, $C_4(z)$, $C_2(z)$, $S_4(z)$, $C_2'(x)$, $C_2'(y)$, $2C_2''$, i, σ_h, $\sigma_v(xz)$, $\sigma_v(yz)$, $2\sigma_d$

9.4

E, $4C_3$, $4S_6$, $C_4(x)$, $C_4(y)$, $C_4(z)$, $3S_4$, $C_2(x)$, $C_2(y)$, $C_2(z)$, $6C_2'$, i, $3\sigma_h$, $6\sigma_d$

9.5

E, C_2

9.6

E, $C_3(z)$, $S_3(z)$, $3C_2$, σ_h, $3\sigma_v$

9.7

E

9.8

$E, C_4, C_2, 2\sigma_v, 2\sigma_d$

9.9

$E, C_3, 3C_2, i, 3\sigma_d$

9.10

$E, C_2(z), C_2(y), C_2(x), i, \sigma(xy), \sigma(xz), \sigma(yz)$

9.11

$E, C_\infty, S_\infty, \sigma_h, i, \infty\sigma_v, \infty C_2$

9.12

$E, C_2(z), C_2(y), C_2(x), i, \sigma(xy), \sigma(xz), \sigma(yz)$

9.13

$E, S_4(z), C_2(z), C_2'(x), C_2'(y), 2\sigma_d$

9.14

$E, C_2(z), \sigma_v(xz), \sigma_v'(yz)$

9.15

$E, C_2(z), C_2(y), C_2(x), i, \sigma(xy), \sigma(xz), \sigma(yz)$

9.16

$E, 4C_3, S_4(x), S_4(y), S_4(z), C_2(x), C_2(y), C_2(z), 6\sigma_d$

9.17

$E, C_3\ (z), S_3\ (Z), 3C_2, \sigma_h, 3\sigma_v$

9.18

E, i

9.19

$E, C_\infty, \infty\sigma_v$

9.20

$E, C_3, 3C_2, i, 3\sigma_d$

9.23

O_h $E, 4C_3, 4S_6, 3C_4, 3S_4, 3C_2, 6C_2', i, 3\sigma_h, 6\sigma_d$

9.24

D_{2d} $E, S_4, C_2, 2C_2', 2\sigma_d$

9.25

C_s E, σ_h

9.26

D_{5h} $E, C_5, S_5, 5C_2, \sigma_h, 5\sigma_v$

9.27

C_{2v} $E, C_2, \sigma_v, \sigma'_v$

9.28

D_{6h} E, C_6, C_3, S_6, S_3, C_2, $3C_2'$, $3C_2''$, i, σ_h, $3\sigma_v$, $3\sigma_d$

9.29

D_{4d} E, S_8, C_4, C_2, $4C_2'$, $4\sigma_d$

9.30

O_h See 9.23

9.31

D_{3h} E, C_3, S_3, $3C_2$, σ_h, $3\sigma_v$

9.32

T_d E, $4C_3$, $3S_4$, $3C_2$, $6\sigma_d$

9.33

C_{2v} See 9.27

9.34

C_s See 9.25

9.35

D_{3h} See 9.31

9.36

C_3 E, C_3

9.37

C_{2v} See 9.27

9.38

C_{3v} E, C_3, $3\sigma_v$

9.39

T_d See 9.32

9.40

C_s See 9.25

9.41

D_{4h} E, C_4, C_2, S_4, $2C_2'$, $2C_2''$, i, σ_h, $2\sigma_v$, $2\sigma_d$

9.42

D_{2d} See 9.24

CHAPTER 10. Molecular Electronic Structure

10.1 Given the equilibrium dissociation energy D^e for N_2 in Table 10.2 and the fundamental vibration frequency 2331 cm^{-1}, calculate the spectroscopic dissociation energy D^o in Kcal mol^{-1}.

SOLUTION:

$$D^e = D^o + \frac{1}{2} h\nu_o$$

$$\frac{1}{2} h\nu_o = \frac{(6.626\times10^{-34}\ Js)(2.998\times10^{8}\ ms^{-1})(2.331\times10^{5}\ m^{-1})}{2(4.184\ J\ cal^{-1})(1000\ cal\ Kcal^{-1})} \times \frac{(6.022\times10^{23}\ mol^{-1})}{1}$$

$$= 3.33\ Kcal\ mol^{-1}$$

$$D^o = D^e - \frac{1}{2} h\nu_o$$

$$= 225.02\ Kcal\ mol^{-1}$$

This is the thermodynamic dissociation energy at absolute zero. The standard change in enthalpy for dissociation at 25 °C calculated from Table A.1 is 225.958 kcal mol^{-1}.

10.2 Calculate the normalization factors shown in equation 10.31.

SOLUTION

$$\psi = C\left[(1s_A) - (1s_B)\right]$$

$$\int|\psi|^2\, d\tau = 1 = C^2 \int\left[(1s_A)^2 - 2(1s_A)(1s_B) + (1s_B)^2\right] d\tau$$

$$1 = C^2\,(1 - 2S_{AB} + 1)$$

$$C = \frac{\pm 1}{\left[2(1 - S_{AB})\right]^{1/2}}$$

10.3 Derive the expression for the normalizi-ation constant C in

$$\psi = C\left[(1s_a) + (1s_b)\right]$$

SOLUTION

$$C^2 \int\left[(1s_a) + (1s_b)\right]^2 d\tau = 1$$

$$= C^2\left[\int(1s_a)^2 d\tau + 2\int(1s_a)(1s_b)\, d\tau + \int(1s_b)^2 d\tau\right]$$

$= C^2\,[1 + 2S + 1]$ where S is the overlap integral

$$C = \frac{\pm 1}{\sqrt{2\,(1+S)}}$$

10.5 Using data in Table A.1 calculate the electronegativities of Cl, Br, and I. The electronegativity of H is taken to be 2.1.

SOLUTION:

$$|x_A - x_B| = 0.208\left\{E(A-B) - \frac{1}{2}\left[E(A-A) + E(B-B)\right]\right\}^{1/2}$$

For HCl

$$|x_{Cl} - x_H| = 0.208\left\{22.062 + 52.095 + 29.082 - \frac{1}{2}\left[2(52.095) + 2(29.082)\right]\right\}^{1/2}$$

$$x_{Cl} - 2.1 = 1.0$$

$$x_{Cl} = 3.1$$

For HBr

$$|x_{Br} - x_H| = 0.208\left\{8.70 + 52.095 + 26.741 - \frac{1}{2}\left[2(52.095) + 2 \times (26.741) - 7.387\right]\right\}^{1/2}$$

$$x_{Br} - 2.1 = 0.7$$

$$x_{Br} = 2.8$$

For HI

$$|x_I - x_H| = 0.208\left\{-6.33 + 52.095 + 25.535 - \frac{1}{2}\left[2(52.095)\right.\right.$$

$$+2(25.535) - 14.923\Big]\Big\}^{1/2}$$

$$x_I - 2.1 = 0.2$$

$$x_I = 2.3$$

10.6 For KF (g) the dissociation constant D^e is 5.18 eV and the dipole moment is 8.60 D. Estimate these values assuming that the bonding is entirely ionic. The ionization potential K(g) is 4.34 eV, and the electron affinity of F(g) is 3.40 eV. The equilibrium internuclear distance in KF (g) is 217 pm.

SOLUTION

The work required to separate $K^+ F^-$ from the equilibrium internuclear distance to infinity is

$$\phi = \frac{Q_1Q_2}{4\pi\epsilon_0 R_e} = \frac{(1.602\times10^{-19}\,C)^2(0.8988\times10^{10}\,N\,m^2\,C^{-2})}{217\times10^{-12}\,m}$$

$$= 1.063\times10^{-18}\,J$$

$$= \frac{1.063\times10^{-18}\,J}{1.602\times10^{-19}\,J\,eV^{-1}} = 6.64\,eV$$

Since KF actually dissociates into neutral atoms the ionization potential of the metal has to be subtracted and the electron affinity of the nonmetal has to be added to calculate D^e:

$$D^e = 6.64 - 4.34 + 3.40 = 5.70\,eV$$

This is higher than the experimental

value (5.18 eV) because there is some repulsion of K^+ and F^- due to inner electron clouds.

The dipole moment expected for the oversimplified structure $K^+ F^-$ is

$$\frac{(1.602 \times 10^{-19}\ \mathrm{C})(217 \times 10^{-12}\ \mathrm{m})}{3.334 \times 10^{-30}\ \mathrm{C\ m}} = 10.4 \text{ Debye units}$$

10.8 The molar polarization $\mathcal{P}$ of ammonia varies with temperature as follows:

$t/°\mathrm{C}$	19.1	35.9	59.9	113.9	139.9	172.9
$\mathcal{P}/\mathrm{cm^3\ mol^{-1}}$	57.57	55.01	51.22	44.99	42.51	39.59

Calculate the dipole moment of amonia.

SOLUTION

$$\mathcal{P} = \frac{N_A}{3\epsilon_0}\left(\frac{\mu^2}{3kT} + \alpha_e + \alpha_v\right)$$

The slope of a plot of $\mathcal{P}$ versus $1/T$ is

$$\frac{N_A \mu^2}{9k\epsilon_0} = \frac{(66-5)\ \mathrm{cm^3\ mol^{-1}}}{4 \times 10^{-3}\ \mathrm{K^{-1}}}$$

$$= 1.525 \times 10^4\ \mathrm{cm^3\ mol^{-1}\ K}$$

$$= 1.525 \times 10^{-2}\ \mathrm{m^3\ mol^{-1}\ K}$$

$$\mu = \left[\frac{(1.525 \times 10^{-2}\ \mathrm{m^3\ mol^{-1}\ K})\, 9k\epsilon_0}{N_A}\right]^{1/2}$$

$$= \left[\frac{(1.525\times10^{-2}\,m^3\,mol^{-1}\,K)(9)(1.381\times10^{-23}\,J\,K^{-1})(8.854\times10^{-12}\,C^2\,N^{-1}\,m^{-2})}{6.022\times10^{23}\,mol^{-1}}\right]$$

$= 5.28 \times 10^{-30}$ C m

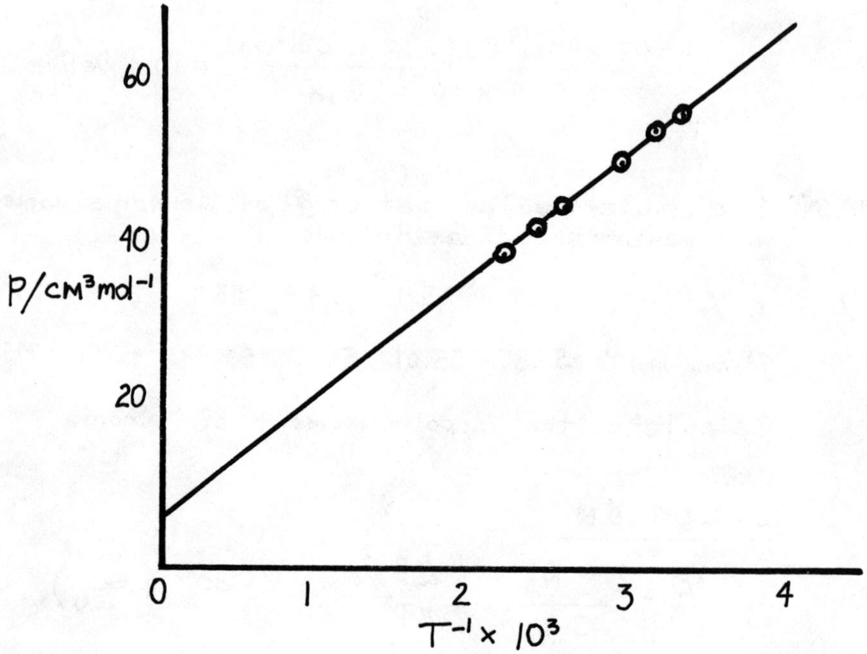

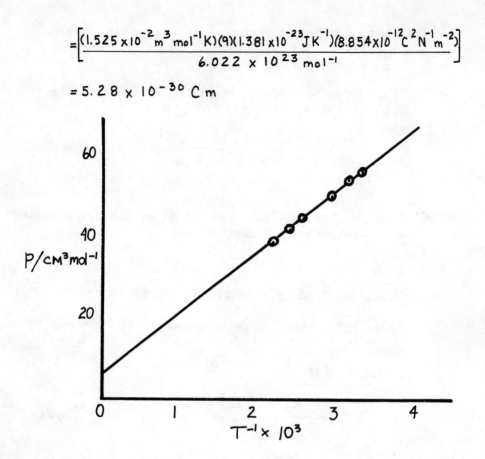

10.9 Show that the SI units are the same on the two sides of equation 10.52.

SOLUTION

$$\frac{K_0-1}{K_0+2}\,\frac{M}{\rho} = \frac{N_A}{3\epsilon_0}\left(\frac{\mu^2}{3kT} + \alpha_e + \alpha_v\right)$$

μ = C m

$$\alpha = \frac{\text{dipole moment}}{\text{electric field strength}} = \frac{C\,m}{V\,m^{-1}} = \frac{C\,m^2}{J\,C^{-1}} = \frac{C^2\,m^2}{kg\,m^2\,s^{-2}}$$

$\epsilon_0 = C^2\,N^{-1}\,m^{-2}$

Left side : $\frac{\text{kg mol}^{-1}}{\text{kg m}^{-3}} = \text{m}^3 \text{ mol}^{-1}$

Right side : $\frac{\text{mol}^{-1}}{\text{C}^2 \text{N}^{-1} \text{m}^{-2}} \left(\frac{\text{C}^2 \text{ m}^2}{\text{kg m}^2 \text{ s}^{-2}} + \frac{\text{C}^2 \text{ m}^2}{\text{kg m}^2 \text{ s}^{-2}} \right)$

$= \text{m}^3 \text{ mol}^{-1}$

10.10 A 1atm the dielectric constant of NH_3 gas is 1.00720 at 292.2 K and 1.00324 at 446.0 K. Calculate the dipole moment μ and the polarizability α.

SOLUTION:

$$(\kappa - 1)\ M/\rho = \frac{N_A}{\epsilon_0} \left(\frac{\mu^2}{3kT} + \alpha \right)$$

Assuming NH_3 is a perfect gas $P = \rho RT/M$

At T = 292.2 K,

$M/\rho = RT/P = (0.08206 \text{ L atm K}^{-1} \text{mol}^{-1})(292.2 \text{K})/(1 \text{atm})$

$= 23.98 \text{ L mol}^{-1}$

$= 23.98 \times 10^{-3} \text{ m}^3 \text{ mol}^{-1}$

$(\kappa - 1)\ M/\rho = (1.00720 - 1)(23.98 \times 10^{-3} \text{ m}^3 \text{ mol}^{-1})$

$= 1.727 \times 10^{-4} \text{ m}^3 \text{ mol}^{-1}$

At T = 446.0 K

$M/\rho = (0.08206 \text{ L atm K}^{-1} \text{ mol}^{-1})(446.0 \text{K})/(1 \text{atm})$

$= 36.60 \text{ L mol}^{-1}$

$= 36.60 \times 10^{-3} \text{ m}^3 \text{ mol}^{-1}$

$$(\kappa - 1) M/\rho = (1.00324 - 1)(36.60 \times 10^{-3}\ m^3\ mol^{-1})$$

$$= 1.186 \times 10^{-4}\ m^3\ mol^{-1}$$

We have the following two simultaneous equations:

$$1.727 \times 10^{-4}\ m^3\ mol^{-1} = \frac{N_A \mu^2}{3\epsilon_0 k(292.2K)} + \frac{N_A \alpha}{\epsilon_0}$$

$$1.186 \times 10^{-4}\ m^3\ mol^{-1} = \frac{N_A \mu^2}{3\epsilon_0 k(446.0K)} + \frac{N_A \alpha}{\epsilon_0}$$

Taking the difference

$$0.541 \times 10^{-4}\ m^3\ mol^{-1} = \frac{N_A \mu^2}{3\epsilon_0 k}\left(\frac{1}{292.2K} - \frac{1}{446.0K}\right)$$

$$\frac{N_A \mu^2}{3\epsilon_0 k} = 0.0458\ m^3\ mol^{-1}\ K$$

$$\mu = \left[\frac{(0.0458\ m^3 mol^{-1} K)(3)(8.854 \times 10^{-12} C^2 N^{-1} m^{-2})(1.3807 \times 10^{-23}}{6.022 \times 10^{23}\ mol^{-1}} \times \frac{JK^{-1})}{1}\right]^{1/2}$$

$$= 5.28 \times 10^{-30}\ Cm$$

Solving the first equation for α

$$\alpha = \frac{\epsilon_0}{N_A}\left[1.727 \times 10^{-4} m^3 mol^{-1} - \frac{N_A \mu^2}{3\epsilon_0 k(292.2K)}\right]$$

$$= \frac{8.854 \times 10^{-12} C^2 N^{-1} m^{-2}}{6.022 \times 10^{23}\ mol^{-1}}\left[1.727 \times 10^{-4} m^3 mol^{-1} - \right.$$

$$\left.\frac{(6.022 \times 10^{23} mol^{-1})(5.28 \times 10^{-30} C m)^2}{3(8.854 \times 10^{-12} C^2 N^{-1} m^{-2})(1.3807 \times 10^{-23} JK^{-1})(292.2K)}\right]$$

$= 2.36 \times 10^{-40}$ Cm/Vm^{-1}

10.11 118 kcal mol^{-1}. 117.966 kcal mol^{-1} from Table A.2.

10.13 $-2\,(13.598\ 396\,\text{eV}) = -27.196\ 792\,\text{eV}$

$-27.196\ 792 - 4.7483 = -31.9451$ eV

10.16 More charge is transferred in forming HCl than in forming HI. This makes HCl more stable than HI. This is in accord with the fact that more heat is evolved when HCl is formed from its elements than HI.

10.17 Because of the higher electronegativity of Cl, the Cl atom in CH_3Cl is more negative than the Br atom in CH_3Br.

10.18 6.408×10^{-29} Cm

10.19 4.0×10^{-30} Cm. 0.88×10^{-40} Cm/Vm^{-1}

10.20 5.29×10^{-30} Cm.

CHAPTER 11 Spectroscopy

11.1 Since the energy of a molecular quantum state is divided by kT in the Boltzmann distribution, it is of interest to calculate the temperature of which kT is equal to the energy of photons of different wavelength. Calculate the temperature at which kT is equal to the energy of photons of wavelength 10^3 cm, 10^{-1} cm, 10^{-3} cm, 10^{-5} cm.

SOLUTION:

$$kT = hc/\lambda$$

$$T = \frac{hc}{k\lambda} = \frac{(6.626 \times 10^{-34} \text{ Js})(2.998 \times 10^{8} \text{ ms}^{-1})}{(1.381 \times 10^{-23} \text{ JK}^{-1})(10 \text{ m})}$$

$$= 1.44 \times 10^{-3} \text{ K} \quad \text{for } \lambda = 10^{3} \text{ cm}$$

For $\lambda = 10^{-1}$ cm, $T = 14.4$ K

For $\lambda = 10^{-3}$ cm, $T = 1440$ K

For $\lambda = 10^{-5}$ cm, $T = 144{,}000$ K

11.2 Most chemical reactions require activation energies ranging between 10 and 100 Kcal mol^{-1}. What are the equivalents of 10 and 100 Kcal mol^{-1} in terms of (a) nm (b) wave numbers, (c) electron volts?

SOLUTION:

For 10 Kcal mol^{-1}

(a) $\lambda = \frac{hc}{E}$

$$= \frac{(6.626\times10^{-34}\,\text{J s})(2.998\times10^{8}\,\text{m s}^{-1})(6.022\times10^{23}\,\text{mol}^{-1})}{(10^{4}\,\text{cal mol}^{-1})(4.184\ \text{J cal}^{-1})}$$

$= 2.859 \times 10^{-6}$ m

$= 2859$ nm

(b) $\tilde{\nu} = \frac{1}{\lambda} = \frac{1}{2.859 \times 10^{-4}\ \text{cm}} = 3498\ \text{cm}^{-1}$

(c) $$\frac{(10^{4}\,\text{cal mol}^{-1})(4.184\ \text{J cal}^{-1})}{(6.022\times10^{23}\,\text{mol}^{-1})(1.602\times10^{-19}\,\text{J eV}^{-1})}$$

$= 0.434$ eV

For 100 Kcal mol^{-1}

(a) $\lambda = (2859\ \text{nm})\dfrac{10^{4}\ \text{cal mol}^{-1}}{10^{5}\ \text{cal mol}^{-1}} = 285.9\ \text{nm}$

(b) $\tilde{\nu} = \frac{1}{\lambda} = \frac{1}{2.859 \times 10^{-5}\ \text{cm}} = 34980\ \text{cm}^{-1}$

(c) $(0.434\ \text{eV})\dfrac{10^{5}\ \text{cal mol}^{-1}}{10^{4}\ \text{cal mol}^{-1}} = 4.34\ \text{eV}$

11.4 Calculate the frequency in wave numbers and the wavelength in cm of the first rotational transition ($J = 0 \longrightarrow 1$) for $D^{35}Cl$.

SOLUTION:

$\tilde{\nu} = 2BJ = \frac{h}{4\pi c I}$ since $J = 1$

The moment of inertia was calculated in the preceeding problem,

$$\tilde{\nu} = \frac{(6.626 \times 10^{-34}\ \mathrm{J\,s})}{4\pi^2 (2.998 \times 10^8\ \mathrm{m\,s^{-1}})(5.141 \times 10^{-47}\ \mathrm{kg\,m^2})}$$

$$= 1089\ \mathrm{m^{-1}}$$

$$= (1089\ \mathrm{m^{-1}})(10^{-2}\ \mathrm{m\,cm^{-1}}) = 10.89\ \mathrm{cm^{-1}}$$

$$\lambda = 1/\tilde{\nu} = 1/(10.89\ \mathrm{cm^{-1}}) = 0.09183\ \mathrm{cm}$$

11.5 The moment of inertia of $^{12}C\,^{16}O$ is $18.75 \times 10^{-47}\ \mathrm{kg\,m^2}$. Calculate the frequencies in wave numbers and the wavelengths in centimeters for the first four lines in the pure rotational spectrum.

SOLUTION:

$$\tilde{\nu} = \frac{2hJ}{8\pi^2 Ic}$$

$$= \frac{2(6.626 \times 10^{-34}\ \mathrm{J\,s})(10^{-2}\ \mathrm{m\,cm^{-1}})J}{8\pi^2 (18.75 \times 10^{-47}\ \mathrm{kg\,m^2})(2.998 \times 10^8\ \mathrm{m\,s^{-1}})}$$

$$= 2.986\ J\ \mathrm{cm^{-1}}$$

J	$\tilde{\nu}/\mathrm{cm^{-1}}$	$\lambda = \tilde{\nu}^{-1}/\mathrm{cm}$
$0 \longrightarrow 1$	2.986	0.3349
$1 \longrightarrow 2$	5.972	0.1675
$2 \longrightarrow 3$	8.957	0.1116
$3 \longrightarrow 4$	11.943	0.0837

11.6 The separation of the pure rotation lines in the spectrum of CO is $3.86\ \mathrm{cm^{-1}}$. Calculate the equilibrium internuclear separation.

SOLUTION:

$$\mu = \frac{(12\times10^{-3}\ \text{kg mol}^{-1})(15.99491\times10^{-3}\ \text{kg mol}^{-1})}{(27.99491\times10^{-3}\ \text{kg mol}^{-1})(6.022045\times10^{23}\text{mol}^{-1})}$$

$$= 1.13852\times10^{-26}\ \text{kg}$$

$$I = \frac{h}{8\pi^2 c\tilde{B}_e} = \frac{6.626\times10^{-34}\ \text{J s}}{8\pi^2(2.998\times10^8\ \text{m s}^{-1})(1.93\ \text{cm}^{-1})(10^2\ \text{cm m}^{-1})}$$

$$= 14.50\times10^{-47}\ \text{kg m}^2$$

$$R_e = (I/\mu)^{1/2} = [(14.50\times10^{-47}\ \text{kg m}^2)/(1.1385\times10^{-26}\ \text{kg})]^{1/2}$$

$$= 113\ \text{pm}$$

11.7 The moment of inertia of $^{16}O{=}^{12}C{=}^{16}O$ is 7.167×10^{-46} kg m^2. (a) Calculate the CO bond length, R_{CO}, in CO_2. (b) Assuming that isotopic substitution does not alter R_{CO}, calculate the moments of inertia of

1. $^{18}O{=}^{12}C{=}^{18}O$ and
2. $^{16}O{=}^{13}C{=}^{16}O$

SOLUTION

(a) Since the CO_2 molecule is symmetrical the carbon atom is on the axis of rotation and does not contribute to the moment of inertia.

$$I = 2mR_{CO}^2$$

$$R_{CO} = (I/2m)^{1/2}$$

$$= \left[\frac{(71.67 \times 10^{-47} \text{ kg m}^2)(6.022045 \times 10^{23} \text{ mol}^{-1})}{2(15.99491 \times 10^{-3} \text{ kg mol}^{-1})}\right]^{1/2}$$

$= 0.1162$ nm

(b) For $^{18}O\ ^{12}C\ ^{18}O$

$$I = 2 m R_{CO}^2$$

$$= \frac{2(17.99916 \times 10^{-3} \text{ kg mol}^{-1})(0.1162 \times 10^{-9} \text{ m})^2}{6.022045 \times 10^{23} \text{ mol}^{-1}}$$

$$= 8.071 \times 10^{-46} \text{ kg m}^2$$

For $^{16}O\ ^{13}C\ ^{16}O$ the moment of inertia is the same as for $^{16}O\ ^{12}C\ ^{16}O$.

11.9 Calculate the zero-point energies of (a) H_2 and (b) Cl_2 in Kilocalories per mole. The fundamental vibration frequencies are to be found in Table 11.4.

SOLUTION:

$$E = \frac{h\nu_0}{2} = \frac{hc\tilde{\nu}_0}{2}$$

$$E_{H_2} = \frac{(6.626 \times 10^{-34} \text{ Js})(2.998 \times 10^8 \text{ ms}^{-1})(4395 \text{ cm}^{-1})}{2(4.184 \text{ J cal}^{-1})(10^3 \text{ cal kcal}^{-1})} \times \frac{(10^2 \text{ cm m}^{-1})}{} \frac{(6.022 \times 10^{23} \text{ mol}^{-1})}{1}$$

$= 6.283$ Kcal mol^{-1}

$$E_{Cl_2} = \frac{(6.283 \text{ Kcal mol}^{-1})(564.9 \text{ cm}^{-1})}{(4395 \text{ cm}^{-1})}$$

$$= 0.808 \text{ Kcal mol}^{-1}$$

11.10 (a) What vibrational frequency in wave numbers corresponds to a thermal energy of kT at 25°C? (b) What is the wavelength of this radiation?

SOLUTION:

$$hc\tilde{\nu} = kT$$

$$\tilde{\nu} = \frac{kT}{hc}$$

$$= \frac{(1.381 \times 10^{-23} \text{ JK}^{-1})(298.15 \text{ K})(10^{-2} \text{ m cm}^{-1})}{(6.626 \times 10^{-34} \text{ JK}^{-1})(2.998 \times 10^{8} \text{ ms}^{-1})}$$

$$= 207 \text{ cm}^{-1}$$

$$\lambda = \frac{1}{\tilde{\nu}} = 4.83 \times 10^{-3} \text{ cm}$$

11.11 Given the following fundamental vibrational frequencies:

$H^{35}Cl$	2989 cm^{-1}	H^2D	3817 cm^{-1}
$^2D^{35}Cl$	2144 cm^{-1}	$^2D^2D$	2990 cm^{-1}

Calculate $\Delta H°$ for the reaction

$$H^{35}Cl(v=0) + {}^2D\,{}^2D(v=0) = {}^2D\,{}^{35}D\,Cl(v=0) + H^2D(v=0)$$

SOLUTION:

$$\Delta H^\circ = E_{DCl} + E_{HD} - E_{HCl} - E_{D_2}$$

$$E = hc\tilde{\nu}/2 \quad \text{for } v = 0$$

$$\Delta\tilde{\nu} = 2144 + 3817 - 2989 - 2990 = -18 \text{ cm}^{-1}$$

$$\Delta H^\circ = -\frac{1}{2} hc\Delta\tilde{\nu}$$

$$= -\frac{1}{2}(6.63 \times 10^{-34} \text{ J s})(3 \times 10^8 \text{ m s}^{-1})(-18 \text{ cm}^{-1})(100 \text{ cm m}^{-1})$$

$$= 1.79 \times 10^{-22} \text{ J}$$

$$= (1.79 \times 10^{-22} \text{ J})(6.022 \times 10^{23} \text{ mol}^{-1})/(4.184 \text{ J cal}^{-1})$$

$$= 26 \text{ cal mol}^{-1}$$

11.13 Calculate the wavelengths in (a) wave numbers and (b) micrometers of the center two lines in the vibration-rotation spectrum of HBr for the fundamental vibration. The necessary data are to be found in Table 11.4.

SOLUTION:

The reduced mass of $H^{80}Br$ is given in Table 11.5 as

$$\mu = \frac{0.99558 \times 10^{-3} \text{ kg mol}^{-1}}{6.022045 \times 10^{-23} \text{ mol}^{-1}}$$

$$= 1.65323 \times 10^{-27} \text{ kg}$$

$$I = \mu R^2 = (1.65323 \times 10^{-27} \text{ kg})(1.4138 \times 10^{-10} \text{ m})^2$$

$$= 3.30452 \times 10^{-47} \text{ kg m}^2$$

$$\tilde{B} = \frac{h}{8\pi^2 cI}$$

$$= \frac{(6.626 \times 10^{-34}\ \text{Js})(10^{-2}\ \text{m cm}^{-1})}{8\pi^2 (2.998 \times 10^8\ \text{ms}^{-1})(3.3045 \times 10^{-47}\ \text{Kg m}^2)}$$

$$= 8.47\ \text{cm}^{-1}$$

(a) $\tilde{\upsilon}_P = \tilde{\upsilon}_0 - 2BJ''$ where $J'' = 1, 2, 3, \ldots$

$$= 2649.67\ \text{cm}^{-1} - 2(8.47\ \text{cm}^{-1})$$

$$= 2632.73\ \text{cm}^{-1}$$

$\tilde{\upsilon}_R = \tilde{\upsilon}_0 + 2B + 2BJ''$ where $J'' = 0, 1, 2, \ldots$

$$= 2649.67\ \text{cm}^{-1} + 2(8.47\ \text{cm}^{-1})$$

$$= 2666.61\ \text{cm}^{-1}$$

(b) $\lambda_P = 1/\tilde{\upsilon}_P = 3.79834 \times 10^{-4}\ \text{cm}$

$$= 3.79834\ \mu\text{m}$$

$\lambda_R = 1/\tilde{\upsilon}_R = 3.75008 \times 10^{-4}\ \text{cm}$

$$= 3.75008\ \mu\text{m}$$

11.14 The fundamental vibration frequency of $H^{35}Cl$ is $8.667 \times 10^{13}\ s^{-1}$. What would be the separation between the infrared absorption lines for $H^{35}Cl$ and $H^{37}Cl$ if the force constants of the bonds are assumed to be the same?

SOLUTION:

$$\mu = \frac{m_H m_{Cl}}{m_H + m_{Cl}}$$

For $H^{35}Cl$

$$\mu_1 = \frac{(1.007825 \times 10^{-3}\ \mathrm{Kg\ mol^{-1}})(34.96885 \times 10^{-3}\ \mathrm{Kg\ mol^{-1}})}{(35.97668 \times 10^{-3}\ \mathrm{Kg\ mol^{-1}})(6.022045 \times 10^{23}\ \mathrm{mol^{-1}})}$$

$$= 1.62668 \times 10^{-27}\ \mathrm{Kg}$$

For $H^{37}Cl$

$$\mu_2 = \frac{(1.007825 \times 10^{-3}\ \mathrm{Kg\ mol^{-1}})(36.947 \times 10^{-3}\ \mathrm{Kg\ mol^{-1}})}{(37.955 \times 10^{-3}\ \mathrm{Kg\ mol^{-1}})(6.022045 \times 10^{23}\ \mathrm{mol^{-1}})}$$

$$= 1.62911 \times 10^{-27}\ \mathrm{Kg}$$

$$\nu_0 = \frac{1}{2\pi}\left(\frac{k}{\mu}\right)^{1/2}$$

$$\frac{\nu_1}{\nu_2} = \left(\frac{\mu_2}{\mu_1}\right)^{1/2} = \frac{\lambda_2}{\lambda_1}$$

$$\lambda_1 = \frac{(2.9979 \times 10^{8}\ \mathrm{m\ s^{-1}})}{8.667 \times 10^{13}\ \mathrm{s^{-1}}}$$

$$= 3459.0 \times 10^{-9}\ \mathrm{nm}$$

$$\lambda_2 = \lambda_1 \left(\frac{\mu_2}{\mu_1}\right)^{1/2}$$

$$= (3459.0\ \mathrm{nm})\left(\frac{1.62911 \times 10^{-27}\ \mathrm{Kg}}{1.62668 \times 10^{-27}\ \mathrm{Kg}}\right)^{1/2}$$

$$= 3461.6\ \mathrm{nm}$$

$$\lambda_2 - \lambda_1 = 2.6\ \mathrm{nm}$$

11.15 The spectroscopic dissociation energy of H_2 (g) is 4.4763 eV, and the fundamental vibration frequency is 4395.24 cm^{-1}. What is the spectroscopic dissociation energy of D_2 (g) if it has the same force constant?

SOLUTION

Since deuterium atoms are more massive, the fundamental vibration frequency will be lower in D_2 than H_2. Therefore the zero point energy for D_2 is lower, and its dissociation energy is higher.

For H_2 $$\mu = \frac{(1.007825 \times 10^{-3}\ \text{kg mol}^{-1})^2}{2(1.007825 \times 10^{-3}\ \text{kg mol}^{-1})(6.022045 \times 10^{23}\ \text{mol}^{-1})}$$

$$= 8.367797 \times 10^{-28}\ \text{kg}$$

For D_2 $$\mu = \frac{(2.01410 \times 10^{-3}\ \text{kg mol}^{-1})^2}{2(2.01410 \times 10^{-3}\ \text{kg mol}^{-1})(6.022045 \times 10^{23}\ \text{mol}^{-1})}$$

$$= 16.722725 \times 10^{-28}\ \text{kg}$$

According to eqn 11.54

$$\frac{\tilde{\omega}_{D_2}}{\tilde{\omega}_{H_2}} = \left[\frac{8.367797 \times 10^{-28}\ \text{kg}}{16.722725 \times 10^{-28}\ \text{kg}}\right]^{1/2}$$

$\tilde{\omega}_{H_2} = 4395.24\ \text{cm}^{-1}$ so

$\tilde{\omega}_{D_2} = 3109.10\ \text{cm}^{-1}$

$\Delta\tilde{\omega} = 1286.1\ \text{cm}^{-1}$

$$D^\circ_{D_2} = 4.4763\ \text{eV} + hc\Delta\tilde{\omega}/2$$

$$= 4.4763\ \text{eV} + \frac{(6.626 \times 10^{-34}\ \text{J s})(2.9979 \times 10^{8}\ \text{m s}^{-1})}{2(1.602 \times 10^{-19}\ \text{J eV}^{-1})} \times \frac{(1.286 \times 10^{5}\ \text{m}^{-1})}{1}$$

$$= 4.4763 \text{ eV} + 0.0797 \text{ eV}$$

$$= 4.5560 \text{ eV}$$

11.17 List the numbers of translational, rotational, and vibrational degrees of freedom for (a) Ne, (b) N_2, (c) CO_2, and (d) CH_2O.

SOLUTION:

	Molecule	Trans.	Rot.	Vib.	Total = 3N
(a)	Ne	3	0	0	3
(b)	N_2	3	2	1	6
(c)	CO_2	3	2	4*	9
(d)	CH_2O	3	3	6**	12

* For linear molecules 3N - 5

** For nonlinear molecules 3N - 6

11.18 When CCl_4 is irradiated with the 435.8 nm mercury line, Raman lines are obtained at 439.9, 441.8, 444.6, and 450.7 nm. Calculate the Raman frequencies of CCl_4 (expressed in wave numbers). Also calculate the wavelengths (expressed in μm) in the infrared at which absorption might be expected.

SOLUTION:

$$\tilde{\nu}_{Raman} = \frac{1}{\lambda_{incid}} - \frac{1}{\lambda_{scatt}}$$

$$= \frac{1}{435.8 \times 10^{-7} \text{ cm}} - \frac{1}{439.9 \times 10^{-7} \text{ cm}}$$

$$= 214 \text{ cm}^{-1}$$

$$\lambda = \frac{1}{\tilde{\nu}} = \frac{1}{214 \text{ cm}^{-1}} = 46.7 \times 10^{-6} \text{ m}$$

For the remaining lines

$\tilde{\nu}_{Raman}$ /cm^{-1}	312	454	759
λ/μm	32.0	22.0	13.2

11.19 The first several Raman frequencies of $^{14}N_2$ are 19.908, 27.857, 35.812, 43.762, 51.721, and 59.662 cm^{-1}. These lines are due to pure rotational transitions with J = 1, 2, 3, 4, 5, and 6. The spacing between the lines is $4B_e$. What is the internuclear distance?

SOLUTION:

$$\mu = \frac{(14.00307 \times 10^{-3} \text{ kg})^2}{(28.00614 \times 10^{-3} \text{ kg})(6.022045 \times 10^{23} \text{ mol}^{-1})}$$

$$= 1.162651 \times 10^{-26} \text{ kg}$$

The average spacing between lines is 7.951 cm^{-1}, and so $\tilde{B}_e$ = (7.951 cm^{-1})(10^2 cm m^{-1})/4 = 198.78 m^{-1} since $\Delta J = 2$.

$$\tilde{B}_e = \frac{h}{8\pi^2 c\mu R_e^2}$$

$$R_e = \left(\frac{h}{8\pi^2 c\mu \tilde{B}_e}\right)^{1/2}$$

$$= \left[\frac{6.626176 \times 10^{-34}\ \mathrm{J\,s}}{8\pi^2 (2.99792458 \times 10^8\ \mathrm{m s^{-1}})(1.162651 \times 10^{-26}\ \mathrm{kg})(198.78\ \mathrm{m^{-1}})} \right]$$

$= 0.110$ nm

11.21 A solution of dye containing 0.1 g L^{-1} transmits 80% of the light at 435.6 nm in a glass cell 1 cm thick. (a) What percent of light will be absorbed by a solution containing 2 g per 100 cm^3 in a cell 1 cm thick? (b) What concentration will be required to absorb 50% of the light? (c) What percent of the light will be transmitted by a solution of the dye containing 1 g per 100 cm^3 in a cell 5 cm thick? (d) What thickness should the cell be in order to absorb 90% of the light with solution of this concentration?

SOLUTION:

$$\log \frac{I_0}{I} = \epsilon c l$$

$$\log \frac{100}{80} = \epsilon \,(1 \times 10^{-2}\ \mathrm{g\,cm^{-3}})(1\ \mathrm{cm})$$

$$\epsilon = 9.69\ \mathrm{cm^2\,g^{-1}}$$

(a) $$\log \frac{100}{I} = (9.69\ \mathrm{cm^2\,g^{-1}})(2 \times 10^{-2}\ \mathrm{g\,cm^{-3}})(1\ \mathrm{cm})$$

$$\log I = 2 - 0.1938$$

$$I = 64.0\%$$

Alternatively this calculation may be done using exponential notation.

$$I = I_o e^{-2.303 \epsilon c l}$$

$$= (100\%) e^{-2.303 (9.69\ \mathrm{cm^2 g^{-1}})(2 \times 10^{-2}\ \mathrm{g\,cm^{-3}})(1\ \mathrm{cm})}$$

$= 64.0\%$

(b) $\log \dfrac{100}{50} = (9.69\ \text{cm}^2\ \text{g}^{-1})(1\ \text{cm})c$

$C = 0.0311\ \text{g cm}^{-3}$

(c) $\log \dfrac{100}{I} = (9.69\ \text{cm}^2\text{g}^{-1})(10^{-2}\ \text{g cm}^{-3})(5\ \text{cm})$

or

$$I = I_o\, e^{-2.303(9.69\ \text{cm}^2\ \text{g}^{-1})(10^{-2}\ \text{g cm}^{-3})(5\ \text{cm})}$$

$= 32.8\ \%$

(d) $\log \dfrac{100}{10} = (9.69\ \text{cm}^{-2}\text{g}^{-1})(10^{-2}\text{g cm}^{-3})\ell$

$\ell = 10.32\ \text{cm}$

11.23 The following absorption data are obtained for solutions of oxyhemoglobin in pH 7 buffer at 575 nm in a 1-cm cell:

Grams per 100 cm^3	Transmission, %
0.03	53.5
0.05	35.1
0.10	12.3

The molar mass of hemoglobin is 64,000 g mol^{-1}!

(a) Is Beer's law obeyed? What is the molar absorption coefficient? (b) Calculate the percent transmission for a solution containing 0.01 gram per 100 mL.

SOLUTION:

(a)

Grams per 100 cm^3	Mol L^{-1}	I/I_0	$\log(I/I_0)$	ϵ
0.03	4.69×10^{-6}	0.535	-0.272	5.80×10^4
0.05	7.82×10^{-6}	0.351	-0.455	5.82×10^4
0.10	15.64×10^{-6}	0.123	-0.910	5.82×10^4

Beer's law is obeyed and the molar absorption coefficient is 5.81×10^4 $(mol\ L^{-1})^{-1}\ cm^{-1}$.

(b) $\log(I/I_0) = -\epsilon c l$

$$= -[5.81 \times 10^4 (mol\ L^{-1})^{-1} cm^{-1}](1.564 \times 10^{-6} mol\ L^{-1})(1 cm)$$

$$= -0.091$$

$I/I_0 = 0.81$ or 81% transmission

11.24 The protein metmyoglobin and imidazole form a complex in solution. The molar absorption coefficients in L mol^{-1} cm^{-1} of the metmyoglobin (Mb) and the complex (c) are as follows:

λ	$\epsilon_{Mb}/10^3 (mol\ L^{-1})^{-1} cm^{-1}$	$\epsilon_c/10^3 (mol\ L^{-1})^{-1} cm^{-1}$
500 nm	9.42	6.88
630 nm	3.58	1.30

An equilibrium mixture in a cell of 1 cm path length has an absorbance of 0.435 at 500 nm and 0.121 at 650 nm. What are the concentrations of metmyoglobin and complex?

SOLUTION:

$$\log(I_0/I) = A = (\epsilon_1 c_1 + \epsilon_2 c_2) l$$

At 500 nm $\quad 0.435 = 9.42 \times 10^3 C_1 + 6.88 \times 10^3 C_2$

At 630 nm $\quad 0.121 = 3.58 \times 10^3 C_1 + 1.30 \times 10^3 C_2$

Solving these equations simultaneously

$C_1 = 2.17 \times 10^{-5}$ mol L^{-1} metmyoglobin

$C_2 = 3.37 \times 10^{-5}$ mol L^{-1} complex

11.25 (a) Calculate the energy levels for n = 1 and n = 2 for an electron in a potential well of width 0.5 nm with infinite barriers on either side. The energies should be expressed in J and Kcal mol^{-1} (b) If an electron makes a transition from n = 2 to n = 1 what will be the wavelength of the radiation emitted?

SOLUTION:

(a) $$E = \frac{n^2 h^2}{8 m_e a^2}$$

$$= \frac{n^2 (6.626 \times 10^{-34} \text{ J s})^2}{8 (9.110 \times 10^{-31} \text{ Kg})(5 \times 10^{-10} \text{ m})^2}$$

$$= n^2 \, 2.410 \times 10^{-19} \text{ J}$$

$$= 2.410 \times 10^{-19} \text{ J for } n = 1$$

$$= 9.640 \times 10^{-19} \text{ J for } n = 2$$

$$= \frac{n^2 (2.410 \times 10^{-19} \text{J})(6.022 \times 10^{23} \text{ mol}^{-1})}{(4.184 \text{ J cal}^{-1})(10^3 \text{ cal Kcal}^{-1})}$$

$$= n^2 \, 34.69 \text{ Kcal mol}^{-1}$$

$$= 34.69 \text{ Kcal mol}^{-1} \text{ for } n = 1$$

$= 138.76$ Kcal mol^{-1} for $n = 2$

(b) $E_2 - E_1 = \frac{hc}{\lambda}$

$$\lambda = \frac{hc}{E_2 - E_1}$$

$$= \frac{(6.626 \times 10^{-34} Js)(2.998 \times 10^8 ms^{-1})}{(9.640 - 2.410) \times 10^{-19} J}$$

$= 275$ nm

11.26 When α-D-mannose ($[\alpha]_D^{20} = +29.3°$) is dissolved in water, the optical rotation decreases as β-D-mannose is formed until at equilibrium $[\alpha]_D^{20} = +14.2°$. This process is referred to as mutarotation. As expected, when β-D-mannose ($[\alpha]_D^{20} = -17.0°$) is dissolved in water the optical rotation increases until $[\alpha]_D^{20} = +14.2°$ is obtained. Calculate the percentage of α form in the equilibrium mixture.

SOLUTION:

$[\alpha]_\alpha = 29.3°$

$[\alpha]_\beta = -17.0°$

$$[\alpha]_{mixture} = +14.2° = 29.3° f_\alpha - 17.0° f_\beta$$

$$= 29.3° f_\alpha - 17.0° (1 - f_\alpha)$$

$$= -17.0° + 36.3 f_\alpha$$

$f_\alpha = \frac{31.2°}{46.3°} = 0.67 =$ fraction of α form

11.27 (a) 3.10 nm (b) 414 nm

11.28 (a) 1.13852×10^{-26} kg (b) 1.4492×10^{-46} kg m^2

11.29 21.1, 42.2, 63.3, 189.9 cm^{-1}. 474, 237, 158, 53 μm.

11.30 (a) 4.61×10^{-48} (b) 6.15×10^{-48}

(c) 6.92×10^{-48} (d) 9.22×10^{-48} kg m^2

11.31 0.1079 kcal mol^{-1} 0.0265 cm

11.32 (a) 4.37×10^{-47} kg m^2 (b) 0.163 nm

11.33 $\nu = Jh/4\pi^2 I$

11.34 12 168.5, 24 337.0, 36 505.6 Mc

11.35 0.1855, 0.1737 eV

11.36 349.755 cm^{-1}

11.37 8.731×10^{-20} J 12.57 kcal mol^{-1} 0.5449 eV

11.38 4592.8, 4605.9, 4632.1, 4645.2 cm^{-1}

2.1773, 2.1154, 2.1588, 2.1528 μm

11.39 1870 cm^{-1} 8.45 cm^{-1}

11.40 6.216×10^{13} s^{-1} 4.83 μm

11.41 4.09×10^{8} N m^{-1}

11.42 8.9×10^{-7} 0.358

11.43

Molecule	Translational	Rotational	Vibrational
Cl_2	3	2	3N - 5 = 1
H_2O	3	3	3N - 6 = 3
HC≡CH	3	3	3N - 5 = 7

11.44

Molecule	Translational	Rotational	Vibrational
NNO	3	2	3N - 5 = 4
NH_3	3	3	3N - 6 = 6

11.45 0.07519 nm

11.46 -18.6 kcal mol^{-1}

11.47 (a) 15,885 cm^{-1} (b) 1.9695 eV

11.48 25.2%

11.49 $(7.7 \pm 0.7) \times 10^{-5}$ g cm^{-3}

11.50 2.6 μg cm^{-3}

11.51 (a) 3.24×10^{-3} mol L^{-1} 5.12×10^{-3} mol L^{-1}

(b) 2.95×10^{4} (mol $L^{-1})^{-1}$

11.52 0.522 nm

11.53 (a) $\lambda = 8ml^2 ck^2/(2k+1)h$

(b) 207, 589, 1230 nm

11.54 4.68 eV

11.55 0.418

11.56 0.636

CHAPTER 12 Magnetic Resonance

12.1 Calculate the magnetic flux density to give a precessional frequency for fluorine of 60 MHz.

SOLUTION:

$$B = \frac{h\nu}{g_N \mu_N}$$

$$= \frac{(6.6262 \times 10^{-34}\, Js)(6 \times 10^{7}\, s^{-1})}{(5.257)(5.0508 \times 10^{-31}\, J\, G^{-1})}$$

$$= 14{,}973\ G$$

12.2 Frequencies used in nuclear magnetic resonance spectra are of the order of 60 MHz. Calculate the corresponding energy in Kilocalories per mole

SOLUTION:

$$E = h\nu$$

$$= \frac{(6.626 \times 10^{-34}\, Js)(6 \times 10^{7} s^{-1})(6.022 \times 10^{23}\, mol^{-1})}{(4.184\ J\ cal^{-1})(1000\ cal\ Kcal^{-1})}$$

$$= 5.722 \times 10^{-6}\ Kcal\ mol^{-1}$$

12.3 What magnetic flux density is required for proton magnetic resonance at 220 MHz?

$$\frac{220\ \text{MHz}}{60\ \text{MHz}} \times 14{,}092\ \text{G} = 51{,}670\ \text{G}$$

12.4 The gyromagnetic ratio γ_N for a nucleus is defined by

$$\underset{\sim}{\mu}_N = \gamma_N \hbar \underset{\sim}{I}$$

What is the value of γ_N for H?

<u>SOLUTION:</u>

$$\mu_N = g_N \mu_N I$$

Therefore, $\gamma_N = g_N \mu_N / \hbar$

$$= \frac{2\pi(5.585)(5.0508 \times 10^{-31}\ \text{J G}^{-1})}{(6.6262 \times 10^{-34}\ \text{J s})}$$

$$= 26{,}750\ \text{s}^{-1}\ \text{G}^{-1}$$

12.6 In a magnetic field of 20,000G what fraction of the protons have their spin lined up with the field at room temperature?

<u>SOLUTION:</u>

$$\frac{N_\ell}{N_h} = 1 + \frac{g_N \mu_N B}{kT}$$

$$= 1 + \frac{(5.585)(5.05 \times 10^{-31}\ \text{J G}^{-1})(2 \times 10^4\ \text{G})}{(1.38 \times 10^{-23}\ \text{J K}^{-1})(298\ \text{K})}$$

$$= 1.00001372$$

$$\frac{N_\ell}{N_\ell + N_h} = \frac{1}{1 + N_h/N_\ell} = \frac{1}{1 + 1/(N_h/N_\ell)}$$

$$= \frac{1}{1 + 1/1.00001372}$$

$$= 0.50000343$$

12.7 Using information from Tables 12.2 and 12.3 sketch the spectrum you would expect for ethyl acetate ($CH_3CO_2CH_2CH_3$).

SOLUTION:

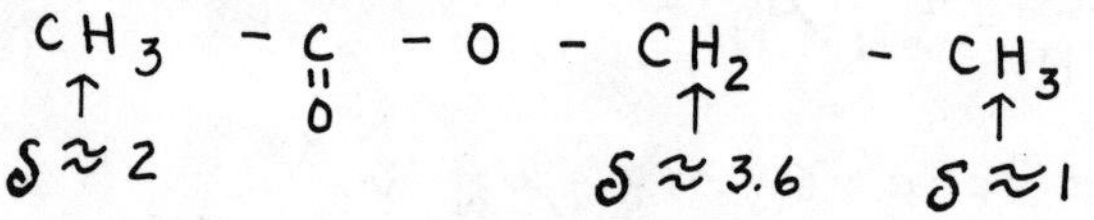

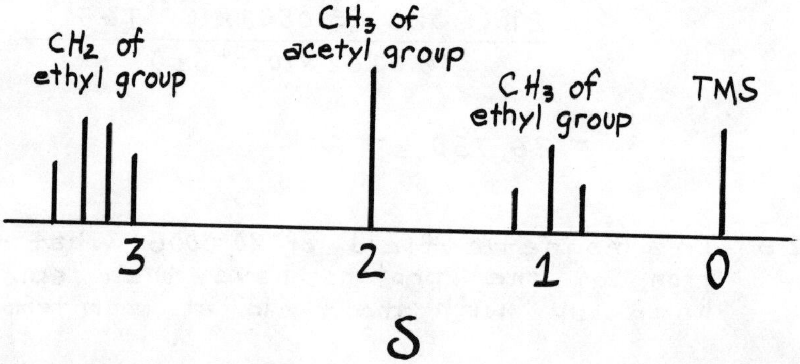

12.10 The proton resonance pattern of 2,3-dibromothiophene shows an AB-type spectrum with lines at 405.22, 410.85, 425.07, and 430.84 Hz measured from tetramethylsilane at 14,100 G (K.F. Kuhlmann and C.L. Braun, J. Chem. Ed., 46, 750 (1969). (a) What is the coupling constant J? (b) What is the difference in the chemical shifts of the A and B hydrogens? (c) At what frequencies would the lines be found at 20,000 G?

SOLUTION:

(a) The average spacing of the two doublets is $J = 5.70 \pm 0.07$ Hz.

(b) $$\nu_0 \delta = \left[(a-d)(b-c)\right]^{1/2}$$

$$= \left[(25.62)(14.22)\right]^{1/2}$$

$$= 19.09 \text{ Hz}$$

$$\delta = (19.09 \text{ Hz})/(60 \times 10^6 \text{ Hz}) = 0.318 \times 10^{-6}$$

$$= 0.318 \text{ ppm}$$

(c) At 20,000 G $\quad \nu_0' = \dfrac{20{,}000}{14{,}100} \; 60 \times 10^6 \text{ Hz}$

$$= 85.1 \times 10^6 \text{ Hz}$$

The center of the spectrum shifts by ν_0' / ν_0

$$418.00 \; \frac{85.1 \times 10^6}{60 \times 10^6} = 592.86$$

Distance of b and c from center =

$$= \frac{\left[(\nu_0 \delta)^2 + J^2\right]^{1/2} - J}{2} = \frac{1}{2}\left\{\left[(85.1 \times 0.318)^2 + 5.70^2\right]^{1/2} - 5.40\right\}$$

$$= 10.98$$

a $= 592.86 - 10.98 - 5.70 \quad = 576.18$ Hz

b $= 592.86 - 10.98 \quad = 581.88$ Hz

c $= 592.86 + 10.98 = 603.84$ Hz

d $= 592.86 + 10.98 + 5.70 = 609.54$ Hz

12.11 Calculate the precessional frequency of electrons in a 15,000-G field.

SOLUTION:

$$\nu = \frac{B g_e \mu_B}{h}$$

$$= \frac{(1.5 \times 10^4 G)(2.0023)(9.2742 \times 10^{-28} J G^{-1})}{6.6262 \times 10^{-34} J s}$$

$$= 4.2037 \times 10^{10} s^{-1}$$

$$= 42{,}037 \text{ MHz}$$

12.12 Line separations in ESR may be expressed in G or MHz. Show how the conversion factor 1G = 2.80 MHz is obtained.

SOLUTION:

The resonance frequency for electrons in a 1 gauss field is given by

$$\nu = \frac{g_e B \mu_B}{h}$$

$$= \frac{(2.00)(9.274 \times 10^{-28} J G^{-1})(1 G)}{6.626 \times 10^{-34} J s}$$

$$= 2.80 \times 10^6 s^{-1}$$

$$= 2.80 \text{ MHz}$$

12.13 Sketch the ESR spectrum expected for p-benzosemiquinone radical ion

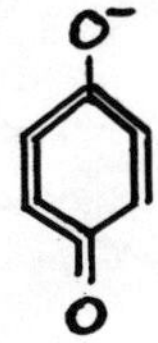

The four hydrogens are magnetically equivalent.

SOLUTION:

1 1	1 H
1 2 1	2 H
1 3 3 1	3 H
1 4 6 4 1	4 H

12.15 Sketch the ESR spectrum for an unpaired electron in the presence of three protons for the following cases: (a) the protons are not equivalent, (b) the protons are equivalent, (c) two protons are equivalent and the third is different.

SOLUTION:

(a)

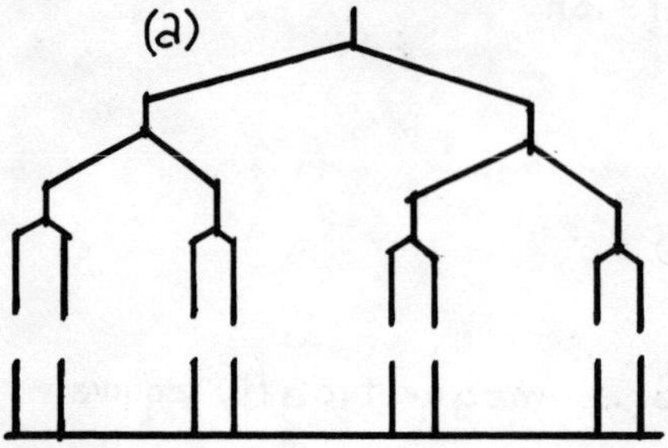

(b)

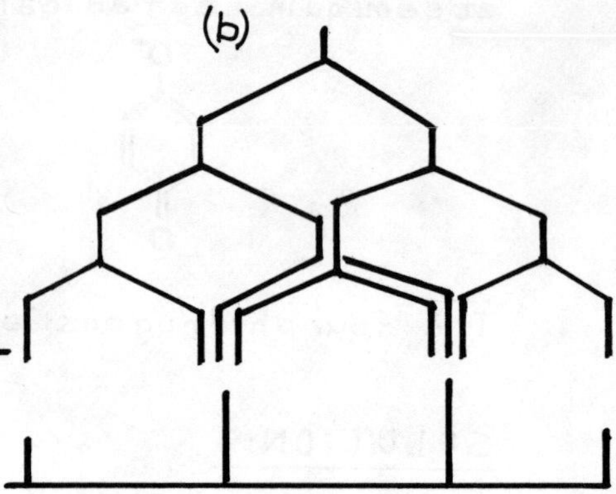

(C)

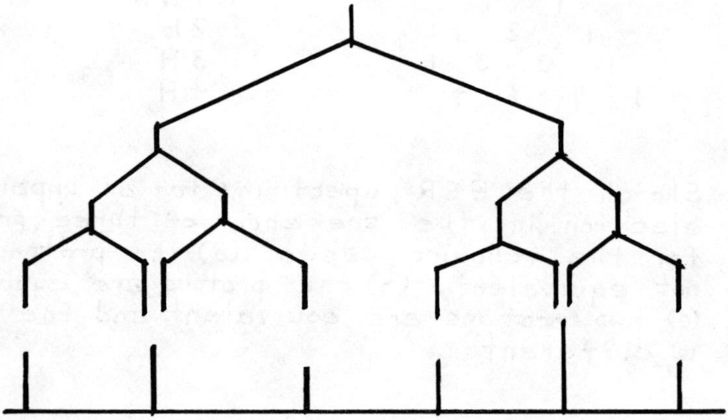

12.16 Sketch the ESR spectrum of CH_3.

SOLUTION:

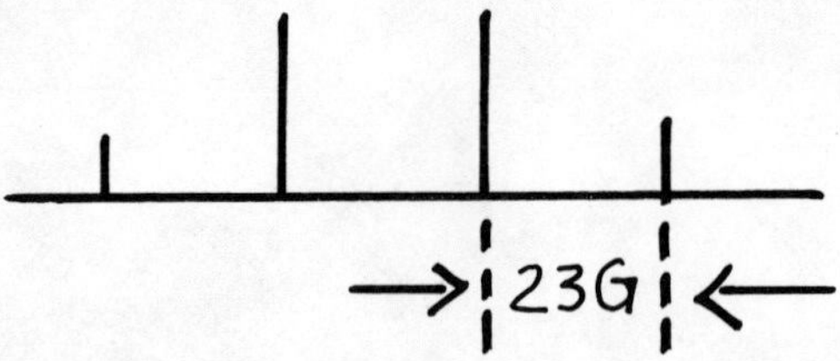

12.17 (a) 4669 G (b) 1248 G

12.18 1.410×10^{-31} J G^{-1}

12.19 4991 G

12.20 10.706 MHz

12.21 0.50000171

12.22 7.270

12.23 Assuming rapid exchange of OH proton

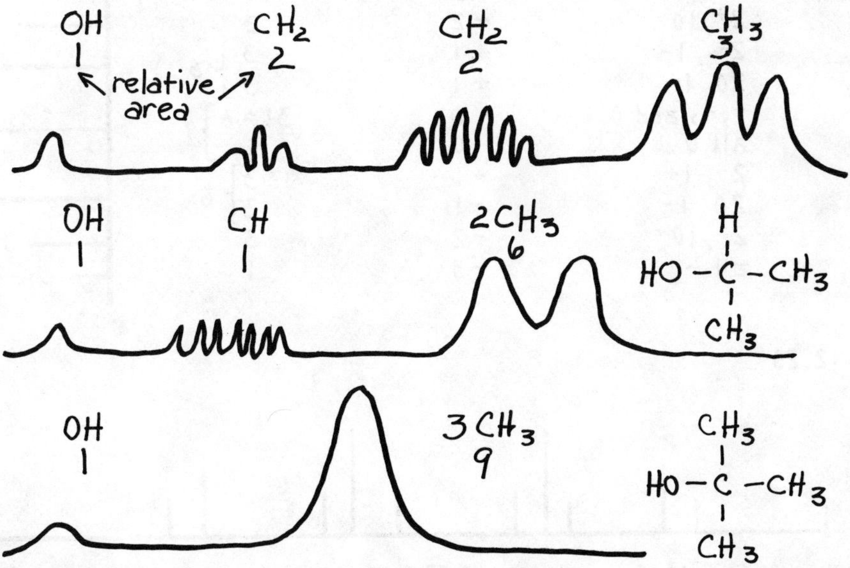

12.24 The proton resonance would be split into three lines by the deuteron and would occur at 42.6 MHz at 10^4 G. The deuteron resonance would be split into two lines by the proton and would occur at 6.5 MHz at 10^4 G.

12.25 The H_2 resonance will be split into a triplet with a coupling constant of 0.5–4 Hz. The H_5 resonance will be split into a triplet with a coupling constant of 6-9 Hz. The H_4 - H_6 resonance will produce an AB-type spectrum.

12.26 9230 MHz

12.27 The deuteron has a spin of +1, 0, or −1. Therefore the following combinations of spins are possible:

Spin comb.	Total spin	No. of ways
all+	+3	1
2^+, 10	+2	3
2^+, 1-	+1	3 } 6
20, 1+	+1	3 } 6
+, −, and 0	0	3! = 6 } 7
all 0	0	1 } 7
2^-, 1+	−1	3 } 6
20, 1−	−1	3 } 6
2^-, 10	−2	3
all −	−3	1

Spectrum

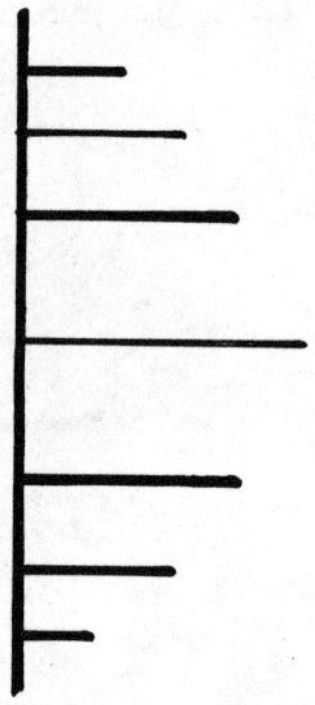

12.28

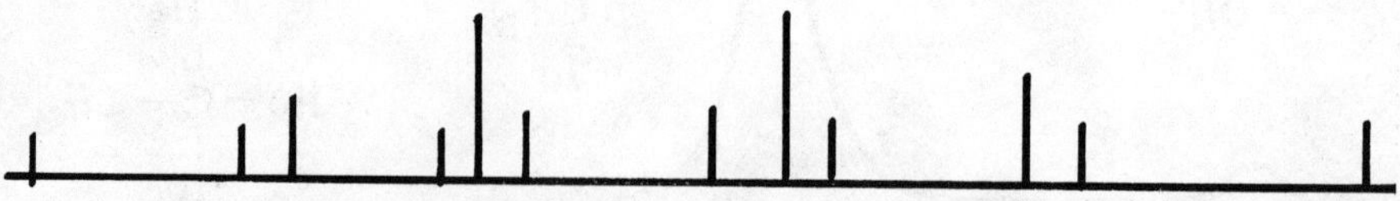

See M. Bersohn and J.C. Baird, An Introduction to Electron Paramagnetic Resonance, Benjamin, New York, 1966, p.93.

12.29

CHAPTER 13 Statistical Mechanics

13.1 (a) How many different ways can two distinguishable balls be placed in two boxes? (b) How many different ways can two distinguishable balls be placed in three boxes? (c) What are the answers to (a) and (b) if the balls are indistinguishable?

SOLUTION:

(a) The number of ways two can be placed in the first box and none in the other is

$W = \frac{N!}{N_1! N_2!} = \frac{2!}{2!\,0!} = 1$. The number of ways one can be placed in each box is $\frac{2!}{1!\,1!} = 2$. The number of ways two can be placed in the second box and none in the other is $\frac{2!}{0!\,2!} = 1$

Thus the total number of different ways that two distinguishable balls can be placed in two boxes is $1+2+1=4$.

(b) $W = \Sigma W_i = \Sigma \frac{N!}{N_1! N_2! N_3!}$

$$= \frac{2!}{1!1!0!} + \frac{2!}{0!1!1!} + \frac{2!}{1!0!1!} + \frac{2!}{2!0!0!}$$

$$+ \frac{2!}{0!2!0!} + \frac{2!}{0!0!2!}$$

$$= 2 + 2 + 2 + 1 + 1 + 1 = 9$$

(c)

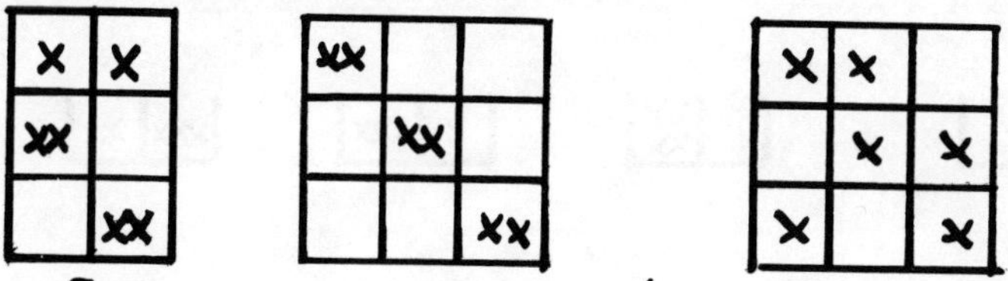

3 6

13.2 (a) How many different ways can four distinguishable balls be placed in two boxes? (b) How many different ways are there if the balls are indistinguishable?

SOLUTION:

(a) $W = \Sigma W_i$

$$W_1 = \frac{4!}{2!\,2!} = 6$$

$$W_2 = \frac{4!}{1!\,3!} = 4$$

$$W_3 = \frac{4!}{3!\,1!} = 4$$

$$W_4 = \frac{4!}{0!\,4!} = 1$$

$$W_5 = \frac{4!}{4!\,0!} = 1$$

$$W = 16$$

(b)

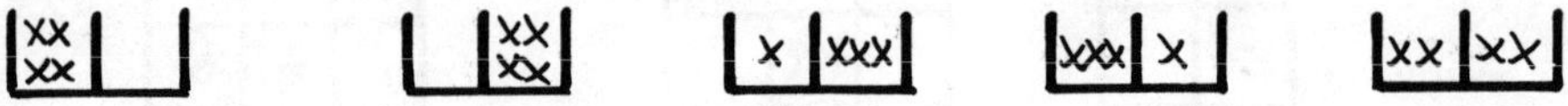

13.3 What are the average occupation numbers N_i for an idealized five-atom crystal with five quanta of energy? As in the example discussed in the chapter we assume that the atoms in the crystal are perfect one-dimensional harmonic oscillators.

SOLUTION:

$$N_i = \frac{5}{\sum_{i=0}^{5} e^{-v_i}} e^{-v_i}$$

$$= \frac{5}{1.578} e^{-v_i}$$

v	0	1	2	3	4	5
N_i	3.169	1.166	0.429	0.158	0.058	0.020

13.4 How many microstates are there for a three-atom crystal of one-dimensional harmonic oscillators if only two quanta of energy are available? On the average how many atoms will have zero, one, or two quanta?

SOLUTION:

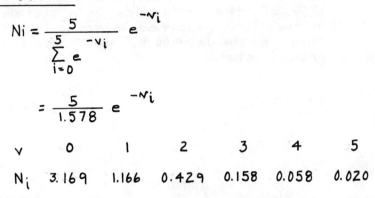

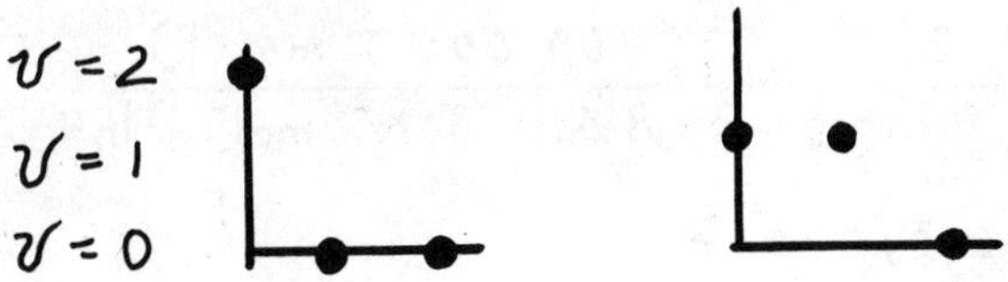

$$W = \frac{3!}{2!\,0!\,1!} = 3 \qquad W = \frac{3!}{1!\,2!\,0!} = 3 \qquad 6 \text{ microstates}$$

$$N_0 = \frac{9}{6} = 1.5$$

$$N_1 = \frac{6}{6} = 1$$

$$N_2 = \frac{3}{6} = 0.5$$

13.6 Calculate the temperature at which 10% of the molecules in a system will be in the first excited electronic state if this state is 400 kJ mol^{-1} above the ground state.

SOLUTION:

$$\frac{N_i}{N} = \frac{e^{-E_i/RT}}{\sum e^{-E_i RT}}$$

$$\frac{N_1}{N} = \frac{e^{-E_1/RT}}{e^{0/RT} + e^{-E_1/RT}}$$

$$0.1 = \frac{1}{e^{E_1/RT} + 1}$$

$$e^{E_1/RT} = \frac{0.9}{0.1} = 9$$

$$\frac{E_1}{RT} = \ln 9$$

$$T = \frac{E_1}{R \ln 9} = \frac{(400,000 \text{ J mol}^{-1})}{(8.314 \text{ J K}^{-1} \text{ mol}^{-1}) \ln 9}$$

$$= 22,000 \text{ K}$$

13.7 Calculate the fraction of hydrogen atoms which at equilibrium at 1000°C would have n=2

SOLUTION:

Since the fraction will be very small, it is given by the ratio of the number with N=2 to the number N=1.

$$\text{Fraction} = \frac{e^{-E_2/RT}}{e^{-E_1/RT}} = e^{-(E_2 - E_1)/kT}$$

From Example 8.5

$$E = \frac{-2.179907 \times 10^{-18}\ J}{n^2}$$

$$\text{Fraction} = \exp\left[-\frac{(-2.179907 \times 10^{-18} J)(-0.75)}{(1.380662 \times 10^{-23} J\ K^{-1})(1273\ K)}\right]$$

$$= 4 \times 10^{-41}$$

13.8 The difference of 1.10 cal $K^{-1} mol^{-1}$ between the third law entropy of CO and the statistical mechanical value can be attributed to the randomness of orientation of CO molecules in its crystals at absolute zero. If half of the molecules are oriented CO and half OC, calculate the entropy of a crystal at absolute zero using equation 13.1 and Stirlings approximation.

SOLUTION:

$$W = \frac{N!}{(N/2)!\,(N/2)!}$$

Stirling's approximation $\ln N! \cong N \ln N - N$

$$\ln W = \ln N! - \ln \frac{N}{2}! - \ln \frac{N}{2}!$$

$$= N \ln N - N - 2\frac{N}{2} \ln \frac{N}{2} + 2\frac{N}{2}$$

$$= N \ln N - N - N \ln \frac{N}{2} + N$$

$$= N \ln 2$$

$$S = k \ln W$$

$$= k N \ln 2$$

$$= (1.987\ \text{cal K}^{-1}\ \text{mol}^{-1})(2.303)(0.301)$$

$$= 1.38\ \text{cal K}^{-1}\ \text{mol}^{-1}$$

13.9 Calculate the translational partition function for H_2 (g) at 1000 K and 1 atm.

SOLUTION:

$$V = \frac{RT}{P} = \frac{(1\ \text{mol})(8.314\ \text{J K}^{-1}\text{mol}^{-1})(1000\ \text{K})}{(101325\ \text{N m}^{-2})}$$

$$= 0.08205\ \text{m}^3$$

$$m = 2\,(1.0078 \times 10^{-3}\ \text{kg mol}^{-1})/(6.022 \times 10^{23}\ \text{mol}^{-1})$$

$$= 3.347 \times 10^{-27}\ \text{kg}$$

$$q_{tr} = \left(\frac{2\pi m kT}{h^2}\right)^{3/2} V$$

$$= \left[\frac{2\pi(3.347 \times 10^{-27}\ \text{kg})(1.3806 \times 10^{-23}\ \text{J K}^{-1})(1000\ \text{K})}{(6.626 \times 10^{-34}\ \text{J s})^2}\right]^{3/2} \times (0.08205\ \text{m}^3)$$

$$= 1.396 \times 10^{30}$$

13.10 Calculate the entropy of one mole of H-atom gas at 1000 K and (a) 1 atm and (b) 1000 atm.

SOLUTION:

$$S^\circ = R \ln\left[\frac{(2\pi m kT)^{3/2}\, V e^{5/2}}{h^3 N_A}\right] + R \ln 2$$

$$m = \frac{1.0079 \times 10^{-3}\ \text{kg mol}^{-1}}{6.022045 \times 10^{23}\ \text{mol}^{-1}}$$

$$= 1.67368 \times 10^{-27}\ \text{kg}$$

$$V = 22.4138 \times 10^{-3}\ \text{m}^3\ \text{mol}^{-1}\ \frac{1000\text{K}}{273.15\text{K}}$$

$$= 0.0820567\ \text{m}^3\ \text{mol}^{-1}\ \text{at 1 atm}$$

$$V = 8.20567 \times 10^{-5}\ \text{m}^3\ \text{mol}^{-1}\ \text{at 1000 atm}$$

$$R = (8.31441\ \text{J K}^{-1}\ \text{mol}^{-1})/(4.184\ \text{J cal}^{-1})$$

$$= 1.98719\ \text{cal K}^{-1}\ \text{mol}^{-1}$$

$$k = 1.380662 \times 10^{-23}\ \text{J K}^{-1}$$

$$T = 1000\ \text{K}$$

$$e = 2.7182818$$

$$h = 6.626176 \times 10^{-34}\ \text{J K}^{-1}$$

$$N_A = 6.022045 \times 10^{23}\ \text{mol}^{-1}$$

(a) At 1000 K and 1 atm

$$S° = 139.761\ \text{J K}^{-1}\ \text{mol}^{-1}$$

$$= 33.404\ \text{cal K}^{-1}\ \text{mol}^{-1}$$

(b) At 1000 K and 1000 atm

$$S° = 82.327\ \text{J K}^{-1}\ \text{mol}^{-1}$$

$$= 19.677\ \text{cal K}^{-1}\ \text{mol}^{-1}$$

13.12 Calculate $S°$ and $C°_P$ for argon ($M = 39.948$ g mol^{-1}) at 25°C and 1 atm

SOLUTION:

$$m = \frac{39.948 \times 10^{-3} \text{ Kg mol}^{-1}}{6.022045 \times 10^{23} \text{ mol}^{-1}}$$

$$= 6.63363 \times 10^{-26} \text{ Kg}$$

$$V = 22.4138 \times 10^{-3} \text{ m}^3 \text{ mol}^{-1} \; \frac{298.15 \text{ K}}{273.15 \text{ K}}$$

$$= 24.4652 \times 10^{-3} \text{ m}^3 \text{ mol}^{-1}$$

$$S^\circ = R\left\{\frac{5}{2} + \ln\left[\left(\frac{2\pi m kT}{h^2}\right)^{3/2} \frac{V}{N_A}\right]\right\}$$

$$= 154.735 \text{ J K}^{-1} \text{ mol}^{-1}$$

$$= 36.983 \text{ cal K}^{-1} \text{ mol}^{-1}$$

$$C^\circ_P = \frac{5}{2} R = 20.786 \text{ J K}^{-1} \text{ mol}^{-1}$$

$$= 4.968 \text{ cal K}^{-1} \text{ mol}^{-1}$$

13.13 Ignoring the vibrational motion of H_2 (g) and any electronic excitation of H (g) or H_2 (g), what value of ΔH_f° does statistical mechanics yield for H(g) at 25°C?

SOLUTION:

At 298.15 K $\frac{1}{2} H_2 \text{(g)} \longrightarrow \text{H (g)}$ $\quad \Delta H^\circ = \Delta H^\circ_{f,H}$

$C_p = \frac{7}{2} R \downarrow$ $\qquad \uparrow C_p \frac{5}{2} R$

At 0K $\frac{1}{2} H_2 \text{(g)} \longrightarrow \text{H (g)}$ $\quad \Delta H^\circ = \frac{(4.4763 \text{ eV})}{2}$

$\times (23.060 \text{ Kcal eV}^{-1})$

$$\Delta H_{f,H}^{\circ} = -\frac{1}{2}\left(\frac{7}{2}\, 1.987192 \text{ cal K}^{-1}\text{ mol}^{-1}\right)(298.15\text{ K}) + \frac{(4.4763\text{ eV})(23.060\text{ Kcal eV}^{-1})}{2} + \frac{5}{2}(1.987192 \text{ cal K}^{-1}\text{ mol}^{-1})(298.15\text{ K})$$

$$= 52.056 \text{ kcal mol}^{-1}$$

[According to Table A.1 $\Delta H_{f,H}^{\circ} = 52.095 \text{ kcal mol}^{-1}$]

13.14 Calculate the translational partition functions for H, H_2, and H_3 at 1000 K and 1 atm. What are the rotational partition functions of H_2 and H_3 (linear) at 1000 K? The internuclear distances in H_3 are 0.094 nm.

SOLUTION:

$$V = \frac{RT}{P} = \frac{(8.31441 \text{ J K}^{-1}\text{mol}^{-1})(1000\text{ K})}{(101325 \text{ N m}^{-2})}$$

$$= 0.082057 \text{ m}^3 \text{ mol}^{-1}$$

$$q_t = \frac{(2\pi m kT)^{3/2} V}{h^3}$$

$$= \frac{[2\pi(1.0079\times10^{-3}\text{ kg}/6022\times10^{23}\text{mol}^{-1})(1.38\times10^{-23}\text{J K}^{-1}) \times (10^3\text{ K})]^{3/2}\, V}{(6.62\times10^{-34}\text{ J s})^3}$$

$$= 6.026\times10^{30}\, V = 4.94\times10^{29}$$

For H_2 $q_t = 6.026\times10^{30}\, V\, 2^{3/2} = 1.40\times10^{30}$

For H_3 $q_t = 6.026\times10^{30}\, V\, 3^{3/2} = 2.57\times10^{30}$

For H_2 $q_r = \frac{8\pi^2 I kT}{2h^2} = 5.72$ (See Example 13.4)

For H_3 $I = \frac{m_1 m_3}{m_1 + m_3} R^2$

$$= \frac{(1.0079 \times 10^{-3}\ \text{kg mol}^{-1})^2\ (1.88 \times 10^{-10}\ \text{m})^2}{2\,(1.0078 \times 10^{-3}\ \text{kg mol}^{-1})(6.02205 \times 10^{23}\ \text{mol}^{-1})}$$

$$= 2.96 \times 10^{-47}\ \text{kg m}^2$$

(Note that m_2 is on the axis of rotation, and does not contribute to the moment of inertia.)

$$q_t = 5.72\ \frac{29.6 \times 10^{-48}\ \text{kg m}^2}{4.60 \times 10^{-48}\ \text{kg m}^2}$$

$$= 36.8$$

13.15 Calculate the ratio of the number of HBr molecules in state v = 2, J = 5 to the number in state v = 1, J = 2 at 1000 K. Assume that all of the molecules are in their electronic ground states. ($\Theta_v = 3700$ K, $\Theta_r = 12.1$ K.)

SOLUTION:

$$\frac{N(v=2, J=5)}{N(v=1, J=2)} = \frac{g_{2\,vib}\, e^{\frac{-\epsilon_{2\,vib}}{kT}}\ g_{5rot}\, e^{\frac{-\epsilon_{5rot}}{kT}}}{g_{1\,vib}\, e^{\frac{-\epsilon_{1vib}}{kT}}\ g_{2rot}\, e^{\frac{-\epsilon_{2\,rot}}{kT}}}$$

g_{vib} = 1 for all v

g_{rot} = $2J+1 = 5$ for all g_{2rot}

= 11 for all g_{5rot}

$\epsilon_{vib} = vh\nu$

Given $\Theta_v = h\nu/k = 3700\text{ K}$ for HBr

$$\epsilon_{vib} = v k \Theta_v$$

$$\frac{\epsilon_{vib}}{kT} = \frac{v\Theta_v}{T}$$

$$\epsilon_{rot} = \frac{J(J+1)h^2}{8\pi^2 I}$$

Given $\Theta_r = \dfrac{h^2}{8\pi^2 I k} = 12.1\text{ K}$ for HBr

$$\epsilon_{rot} = J(J+1)\Theta_r k$$

$$\frac{\epsilon_{rot}}{kT} = \frac{J(J+1)\Theta_r}{T}$$

$$\frac{N(v=2, J=5)}{N(v=1, J=2)} = \frac{e^{-(2)(3.7)} \times 11 \times e^{-5(5+1)(12.1)/10^3}}{e^{-3.7} \times 5 \times e^{-2(2+1)(12.1)/10^3}}$$

$$= 0.0407$$

13.16 The ground state of Cl(g) is fourfold degenerate. The first excited state is 881 cm^{-1} higher in energy and its twofold degenerate. What is the value of the electronic partition function at 25°C? At 1000 K?

SOLUTION:

$$q = g_0 e^{-0} + g_1 e^{-\epsilon_1/kT}$$

$$= 4 + 2e^{-hc\tilde{\nu}/kT}$$

$$= 4 + 2\exp\left[-\frac{(6.626176\times10^{-34}\,\mathrm{J\,s})(2.99792\times10^{8}\,\mathrm{m\,s^{-1}})(8.81\times10^{4}\,\mathrm{m^{-1}})}{(1.380662\times10^{-23}\,\mathrm{J\,K^{-1}})(298.15\,\mathrm{K})}\right]$$

$$= 4 + 2e^{-4.251}$$

$$= 4.0285$$

At 1000 K

$$q = 4 + 2e^{-1.267}$$

$$= 4.5631$$

13.18 Calculate the entropy of nitrogen gas at 25°C and 1 atm pressure. The equilibrium separation of atoms is 0.1095 nm and the vibrational wave number is 2330.7 cm^{-1}.

SOLUTION:

$$m = \frac{2(14.0067\times10^{-3}\,\mathrm{kg\,mol^{-1}})}{6.022045\times10^{23}\,\mathrm{mol^{-1}}} = 4.65181\times10^{-26}\,\mathrm{kg}$$

$$\frac{kT}{P^\circ} = \frac{(1.380662\times10^{-23}\,\mathrm{J\,K^{-1}})(298.15\,\mathrm{K})}{101325\,\mathrm{N\,m^{-2}}}$$

$$= 4.06261\times10^{-26}\,\mathrm{m^3}$$

$$S^\circ_t = R\left\{\frac{5}{2} + \ln\left[\left(\frac{2\pi m kT}{h^2}\right)^{3/2}\frac{kT}{P^\circ}\right]\right\}$$

$$\ln\left[\left(\frac{2\pi(4.65181\times10^{-26}\,\mathrm{kg})(1.380662\times10^{-23}\,\mathrm{J\,K^{-1}})(298.15\,\mathrm{K})}{(6.626176\times10^{-34}\,\mathrm{J\,s})^2}\right)^{3/2} \times (4.06261\times10^{-26}\,\mathrm{m^3})\right]$$

$$= 15.57814$$

$$S_t^\circ = (1.987192 \text{ cal K}^{-1}\text{mol}^{-1})(\tfrac{5}{2} + 15.57814)$$

$$= 35.925 \text{ cal K}^{-1} \text{ mol}^{-1}$$

$$\Theta_r = \frac{h^2}{8\pi^2 I k}$$

$$\mu = \frac{m_N{}^2}{2m_N} = \frac{m_N}{2} = \frac{14.0067\times10^{-3} \text{ kg mol}^{-1}}{2(6.022045\times10^{23} \text{ mol}^{-1})}$$

$$= 1.16295 \times 10^{-26} \text{ kg}$$

$$I = \mu R^2 = (1.16295\times10^{-26} \text{ kg})(1.095\times10^{-10} \text{ m})^2$$

$$= 1.39441\times10^{-46} \text{ kg m}^2$$

$$\Theta_r = \frac{(6.626176\times10^{-34} \text{ J s})^2}{8\pi^2(1.39441\times10^{-46} \text{ kg m}^2)(1.380662\times10^{-23} \text{ J K}^{-1})}$$

$$= 2.88841 \text{ K}$$

$$S_r^\circ = R \ln\left(\frac{eT}{\sigma\Theta_r}\right)$$

$$= (1.987192 \text{ cal K}^{-1}\text{mol}^{-1}) \ln\left[\frac{(2.71828)(298.15 \text{ K})}{(2)(2.88841 \text{ K})}\right]$$

$$= 9.824 \text{ cal K}^{-1} \text{ mol}^{-1}$$

$$x = \frac{h\nu}{kT} = \frac{hc\tilde{\nu}}{kT} = \frac{(6.626176\times10^{-34} \text{ J s})(2.997925\times10^{8} \text{ m s}^{-1}) \times (2.3307\times10^{5} \text{ m}^{-1})}{(1.380662\times10^{-23} \text{ J K}^{-1})(298.15 \text{ K})}$$

$$= 11.247$$

$$S_v^\circ = R\left[\frac{x}{e^x - 1} - \ln(1 - e^{-x})\right]$$

$$= (1.987192\ \text{cal K}^{-1}\text{mol}^{-1})\left[\frac{11.247}{e^{11.247}-1} - \ln(1 - e^{-11.247})\right]$$

$$= 2.9 \times 10^{-4}\ \text{cal K}^{-1}\text{mol}^{-1}$$

$$S^\circ = S_t^\circ + S_r^\circ + S_v^\circ$$

$$= 35.925 + 9.824 + 0$$

$$= 45.749\ \text{cal K}^{-1}\ \text{mol}^{-1}$$

(Table A.1 gives 45.77 cal K^{-1} mol^{-1}.)

13.20 Calculate the value of the equilibrium constant K_p for the dissociation of CO at 2000 K.

$$CO(g) = C(g) + O(g)$$

Spectroscopic data on CO are available in Table 11.2. The degeneracies of the electronic ground states of C(g) and O(g) are both 9. (The value of K_p at 2000 K calculated from data in Table A.2 is 7.427×10^{-22} atm.)

SOLUTION:

For this reaction $\sum \nu_i = 1$. First we calculate the various factors in equation 13.97 separately.

$$\frac{kT}{P} = \frac{(1.380662 \times 10^{-23}\ \text{J K}^{-1})(2000\ \text{K})}{101{,}325\ \text{N m}^{-2}}$$

$$= 2.72521 \times 10^{-25}\ \text{m}^3$$

$$\left(\frac{2\pi kT}{h^2}\right)^{3/2} = \left[\frac{2\pi(1.380662 \times 10^{-23}\ \text{J K}^{-1})(2000\ \text{K})}{(6.626176 \times 10^{-34}\ \text{J s})^2}\right.$$

$$= 2.48403 \times 10^{71}\ Kg^{-3/2}\ m^{3}$$

$$\left(\frac{m_O m_C}{m_{CO}}\right)^{3/2} = \left[\frac{(15.9994)(12.011)(1.6605655 \times 10^{-27}\ Kg)}{(28.0104)}\right]^{3/2}$$

$$= 1.21599 \times 10^{-39}\ Kg^{3/2}$$

In order to calculate the rotational partition function we need to calculate the moment of inertia I_o ($R = 1.1282 \times 10^{-10}$ m).

$$I = \mu R^2 = \frac{m_C m_O R^2}{m_C m_O}$$

$$= \frac{(12.011)(15.9994)(10^{-3}\ Kg\ mol^{-1})(1.1282 \times 10^{-10}\ m)^2}{(28.010)(6.022045 \times 10^{23}\ mol^{-1})}$$

$$= 1{,}45010 \times 10^{-46}\ Kg\ m^2$$

$$\frac{1}{q_{r\ CO}} = \frac{h^2}{8\pi^2 I kT}$$

$$= \frac{(6.626176 \times 10^{-34}\ Js)^2}{8\pi^2 (1.45010 \times 10^{-46}\ Kg\ m^2)(1.380662 \times 10^{-23} JK^{-1})(2000 K)}$$

$$= 1.38874 \times 10^{-3}$$

$$\frac{1}{q_{v\ CO}} = 1 - e^{-h\nu/kT}$$

$$= 1 - e^{-\frac{(6.626176 \times 10^{-34} Js)(2.99792 \times 10^{8} ms^{-1})(21.7021 \times 10^{4} m^{-1})}{(1.380662 \times 10^{-23} JK^{-1})(2000 K)}}$$

$$= 0.790123$$

$$\frac{q_{e\ C}\, q_{e\ O}}{q_{e\ CO}} = \frac{9 \times 9}{1} = 81$$

$$e^{-\Delta\epsilon_o/kT} = e^{-\frac{(11.1008\ ev)(1.60219 \times 10^{-19} Jev^{-1})}{(1.380662 \times 10^{-23} JK^{-1})(2000 K)}}$$

$$= 1.06474 \times 10^{-28}$$

$$K_p = (2.72521 \times 10^{-25} m^3)(2.48403 \times 10^{71} Kg^{-3/2} m^{-3}) \times$$
$$(1.21509 \times 10^{-39} Kg^{3/2})(1.38874 \times 10^{-3})(0.790123) \times$$
$$(81)(1.06474 \times 10^{-28}) = 7.790 \times 10^{-22} \text{ atm}$$

Although the equilibrium constant comes out dimensionless it effectively has the units of atm because one atm has been used as the standard state.

13.21 What fraction of oxygen molecules is dissociated into atoms at 3000 K and 1 atm ignoring electronic contributions $O_2(g) = 2O(g)$?

SOLUTION:

The following factors are needed to make the calculation using equation 13.97

$$\frac{kT}{P^\circ} = \frac{(1.380662 \times 10^{-23} J K^{-1})(3000 K)}{101,325 N m^{-2}}$$

$$= 4.087822 \times 10^{-25} m^3$$

$$\left(\frac{2\pi kT}{h^2}\right)^{3/2} = \left[\frac{2\pi(1.380662 \times 10^{-23} J K^{-1})(3000 K)}{(6.626176 \times 10^{-34} J s)^2}\right]^{3/2}$$

$$= 4.563456 \times 10^{71} kg^{-3/2} m^{-3}$$

$$\left(\frac{m_O^2}{m_{O_2}}\right)^{3/2} = \left[\frac{(15.99491)^2 (1.6605655 \times 10^{-27} kg)}{31.98982}\right]^{3/2}$$

$$= 1.530424 \times 10^{-39} kg^{3/2}$$

$$I = \mu R_e^2$$

$$= \frac{(15.99491 \times 10^{-3}\,\text{kg mol}^{-1})^2\,(0.120740 \times 10^{-9}\,\text{m})^2}{(31.98982 \times 10^{-3}\,\text{kg mol}^{-1})(6.022045 \times 10^{23}\,\text{mol}^{-1})}$$

$$= 1.936021 \times 10^{-46}\,\text{kg m}^2$$

$$\Theta_r = \frac{h^2}{8\pi^2 I k}$$

$$= \frac{(6.62618 \times 10^{-34}\,\text{J s})^2}{8\pi^2(1.936021 \times 10^{-46}\,\text{kg m}^2)(1.38066 \times 10^{-23}\,\text{J K}^{-1})}$$

$$= 2.08037\,\text{K}$$

$$\frac{T}{\sigma_i \Theta_r} = \frac{3000\,\text{K}}{2(2.08037\,\text{K})} = 721.03$$

$$\Theta_v = \frac{hc\tilde{\omega}}{k}$$

$$= \frac{(6.626176 \times 10^{-34}\,\text{J s})(2.997925 \times 10^{8}\,\text{m s}^{-1})(1580.361\,\text{cm}^{-1})}{(1.380\ 662 \times 10^{-23}\,\text{J K}^{-1})} \times (10^2\,\text{cm m}^{-1})$$

$$= 2273.8\,\text{K}$$

$$\frac{1}{1 - e^{-\Theta_v/T}} = \frac{1}{1 - e^{-2273.8/3000}}$$

$$= 1.88194$$

$$\Pi g_{oi}^{\nu_i} = \frac{9^2}{1} = 81$$ (assuming the degeneracy of the oxygen molecule is unity).

$$e^{-\Delta\epsilon_0/kT} = e^{-\frac{(5.080\,\text{eV})(1.601219 \times 10^{-19}\,\text{J eV}^{-1})}{(1.380662 \times 10^{-23}\,\text{J K}^{-1})(3000\,\text{K})}}$$

$$= 2.95908 \times 10^{-9}$$

$K_p = (4.087822 \times 10^{-25}\ m^3)(4.563456 \times 10^{71}\ kg^{-3/2}\ m^{-3}) \times$

$(1.530424 \times 10^{-39}\ kg^{3/2})(721.03)^{-1}(1.88194)^{-1}(81)(2.95908 \times 10^{-9})$

$= 0.0504$ (Note: Table A.1 yields 0.0126)

$$= \frac{4\alpha^2}{1 \alpha^2}$$

$\alpha = 0.112$

13.23 Using equation 13.97 calculate Kp for $H_2(g) = 2H(g)$ at 3000K.

SOLUTION:

$$\frac{kT}{P°} = \frac{(1.380662 \times 10^{-23}\ J\ K^{-1})(3000K)}{101{,}325\ N\ m^{-2}}$$

$$= 4.087815 \times 10^{-25}\ m^3$$

$$\left(\frac{2\pi kT}{h^2}\right)^{3/2} = \left[\frac{2\pi(1.380662 \times 10^{-23}\ J\ K^{-1})(3000\ K)}{(6.626176 \times 10^{-34}\ J\ s)^2}\right]^{3/2}$$

$$= 4.56346 \times 10^{71}\ kg^{-3/2}\ m^{-3}$$

$$\left(\frac{m_H^{\ 2}}{m_{H_2}}\right)^{3/2} = \left[\frac{(1.007825)^2(1.6605655 \times 10^{-27}\ kg)}{2.015650}\right]^{3/2}$$

$$= 2.420565 \times 10^{-41}\ kg^{3/2}$$

$$\frac{T}{\sigma_i \Theta_r} = \frac{3000\ K}{2(87.494K)} = 17.144$$

where the value of Θ_r is given in Example 13.4.

The value of $\left(1-e^{-\Theta_v/T}\right)^{-1}$ is shown to be 1.13753 in Example 13.5.

$$\Pi g_{oi}^{\nu_i} = \frac{2^2}{1} = 4$$

$$e^{-\Delta\epsilon_0/kT} = e^{-\frac{(4.476\ \text{eV})(1.601219\times10^{-19}\ \text{J eV}^{-1})}{(1.380662\times10^{-23}\ \text{J K}^{-1})(3000\ \text{K})}}$$

$$= 3.06\times10^{-8}$$

$$K_P = (4.088\times10^{-25}\ \text{m}^3)(4.563\times10^{71}\ \text{kg}^{-3/2}\ \text{m}^{-3})\times$$

$$(2.421\times10^{-41}\ \text{kg}^{3/2})(17.14)^{-1}(1.138)^{-1}(4)(3.06\times10^{-8})$$

$$= 2.83\times10^{-2}$$

$$= \frac{4\alpha^2}{1-\alpha^2}$$

$$2.83\times10^{-2} - 2.83\times10^{-2}\alpha^2 = 4\alpha^2$$

$$\alpha = \sqrt{\frac{2.83\times10^{-2}}{4.0283}} = 0.0838$$

The experimental value of α obtained by Langmuir is 0.072.

13.24 (a) 8 (b) 4

13.25 $N_0 = 2.86$ $N_1 = 1.43$ $N_2 = 0.57$ $N_3 = 0.14$

13.26 (a) 8 microstates

(b)

$E_1 = 3u$	$E_2 = u$	$E_3 = -u$	$E_4 = -3u$
$N_1 = 1$	$N_2 = 3$	$N_3 = 3$	$N_4 = 1$
$P_1 = 1/8$	$P_2 = 3/8$	$P_3 = 3/8$	$P_4 = 1/8$

(c) $q = e^{u/kT} + e^{-u/kT}$

$Q = e^{3u/kT} + 3e^{u/kT} + 3e^{-u/kT} + e^{-3u/kT}$

(d) $S = 1.734 \times 10^{-23}$ J K^{-1}

$U = 0$

(e) $S = R \ln (e^{u/kT} + e^{-u/kT})$

$U = 0$

13.27 (a) 3.85×10^{-173} (b) 2.47×10^{-23}

13.28 (a) 1.253×10^{-17} (b) 1.147×10^{-169}

(c) 1.100×10^{-4} 2.596×10^{-40}

13.30 $q_{t,I} = 6.98 \times 10^{32}$

13.31 26.016, 33.860, 33.399 cal K^{-1} mol^{-1}

13.32 30.126 cal K^{-1} mol^{-1}

13.33 (a) 3 (b) 4

13.35 2.18×10^{-6}

13.36 (a) 3.369 (b) 8.144

13.37 36.6×10^{36}

13.38 29.060 cal K^{-1} mol^{-1}

13.39 60.165 cal K^{-1} mol^{-1}

13.40 53.265 cal K^{-1} mol^{-1}

13.41 5.34×10^{-3}

13.44 77.3

13.45 3.26

13.46 12.900 cal K^{-1} mol^{-1} The classical expectation is 14.903 cal K^{-1} mol^{-1}

13.47 (a) 6.955 (b) 7.896 cal K^{-1} mol^{-1}

13.48 $B = 2\pi N_A d^3/3$

PART THREE: DYNAMICS

CHAPTER 14 Kinetic Theory of Gases

14.1 If the diameter of a gas molecule is 0.4 nm, and each is imagined to be in a separate cube, what is the length of the side of the cube in molecular diameters at 0°C and pressures of (a) 1 atm and (b) 10^{-3} torr.

SOLUTION:

(a) $$\left[\frac{(22.4\text{ L mol}^{-1})(1000\text{ cm}^3\text{ L}^{-1})(10^7\text{ nm cm}^{-1})^3}{6.02\times10^{23}\text{ mol}^{-1}}\right]^{1/3}$$

$$= 3.338\text{ nm}$$

$$= \frac{3.338\text{ nm}}{0.4\text{ nm}} = 8.3\text{ molecular diameters}$$

(b) $$\left[\frac{(22.4\text{ L atm mol}^{-1})(760\text{ torr atm}^{-1})(10^{-3}\text{ torr})^{-1}(10^{24}\text{ nm}^3\text{ L}^{-1})}{6.02\times10^{23}\text{ mol}^{-1}}\right]^{1/3}$$

$$= 304.7\text{ nm}$$

$$= \frac{304.7\text{ nm}}{0.4\text{ nm}} = 762\text{ molecular diameters}$$

14.2 Assuming that the atmosphere is isothermal at 0°C and that the average molar mass of air is 29 g mol^{-1}, calculate the atmospheric pressure at 20,000 ft. above sea level, using the Boltzmann distribution.

SOLUTION:

The altitude is

$$(20{,}000\ \text{ft.})(12\ \text{in ft}^{-1})(2.54\ \text{cm in}^{-1})(10^{-2}\ \text{m cm})$$

$$= 6096\ \text{m}$$

$$P = P_o\ e^{-Mgh/RT}$$

$$= (1\ \text{atm})\, e^{-\frac{(29\ \text{g mol}^{-1})(10^{-3}\ \text{Kg g}^{-1})(9.80\ \text{ms}^{-2})(6096\ \text{m})}{(8.31\ \text{J K}^{-1}\text{mol}^{-1})(273\ \text{K})}}$$

$$= 0.466\ \text{atm}$$

14.4 Plot the probability density f (v) of various molecular speeds for oxygen at 25°C.

SOLUTION:

$$f(v) = 4\pi v^2 \left(\frac{M}{2\pi RT}\right)^{3/2} e^{-Mv^2/2RT}$$

$$= 4\pi v^2 \left(\frac{32\times10^{-3}\ \text{Kg mol}^{-1}}{2\pi(8.314\text{JK}^{-1}\text{mol}^{-1})(298\text{K})}\right)^{3/2} e^{-\frac{(32\times10^{-3}\ \text{Kg mol}^{-1})v^2}{2(8.314\text{JK}^{-1}\text{mol}^{-1})(298\text{K})}}$$

$$= v^2\ (3.704\times10^{-8}\ \text{s m}^{-1})\, e^{-6.458\times10^{-6}v^2}$$

where v is in ms^{-1}

$v/10^2\ \text{ms}^{-1}$	$f(v)/10^{-4}\ \text{s m}^{-1}$
1	3.47
3	18.64
5	18.42
7	7.67
10	0.58

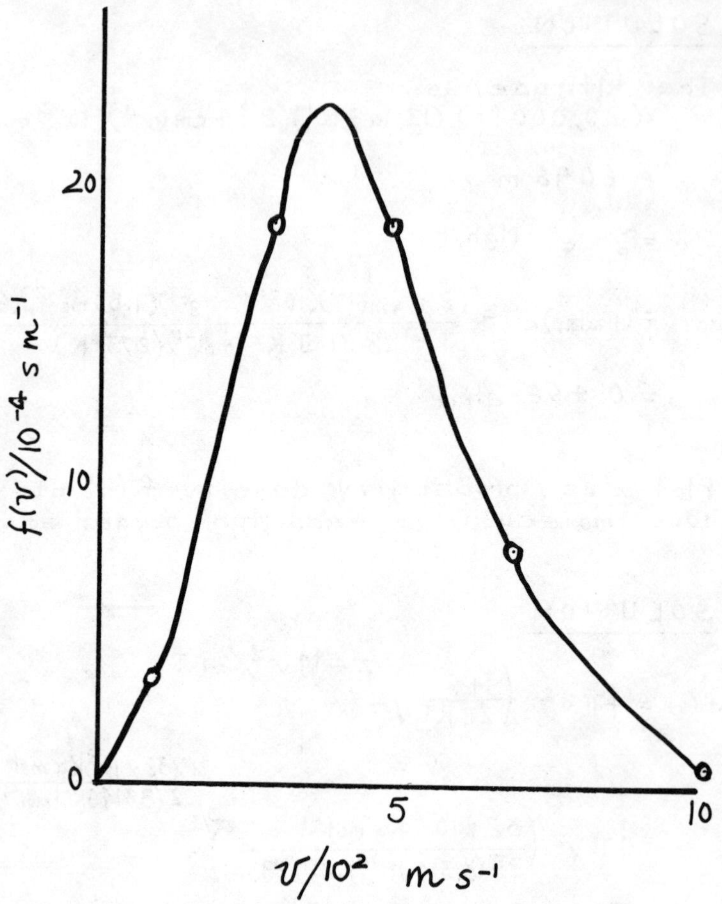

14.5 Calculate the mean speed and the root-mean square speed for the following set of molecules : 10 molecules moving at 5×10^2 ms^{-1}, 20 molecules moving 10×10^2 ms^{-1}, and 5 molecules moving 15×10^2 ms^{-1}.

SOLUTION

$$\langle v \rangle = \frac{\Sigma N_i v_i}{\Sigma N_i} = \frac{10(500) + 20(1000) + 5(1500)}{35}$$

$$= 928 \text{ ms}^{-1}$$

$$\langle v^2 \rangle^{1/2} = \left(\frac{\Sigma N_i v_i^2}{\Sigma N_i} \right)^{1/2} = \left[\frac{10(500)^2 + 20(1000)^2 + 5(1500)^2}{35} \right]^{1/2}$$

$$= 982 \text{ ms}^{-1}$$

14.7 Calculate the velocity of sound in nitrogen gas at 25° C. (See Section 14.4)

SOLUTION:

$$v = \left(\frac{C_P RT}{C_V M} \right)^{1/2}$$

$$= \left[\frac{(6.961 \text{ cal K}^{-1} \text{ mol}^{-1})(8.3144 \text{ JK}^{-1} \text{mol}^{-1})(298.15 \text{K})}{(4.974 \text{ cal K}^{-1} \text{mol}^{-1})(2)(14.0067 \times 10^{-3})} \right]^{1/2}$$

$$= 352 \text{ m s}^{-1}$$

$$= \frac{(352 \text{ m s}^{-1})(10^2 \text{ cm m}^{-1})(60 \text{ s min}^{-1})(60 \text{ min hr}^{-1})}{(2.54 \text{ cm in}^{-1})(12 \text{ in ft}^{-1})(5280 \text{ ft mile}^{-1})}$$

$$= 787 \text{ miles hr}^{-1}$$

14.8 Calculate the number of collisions per square centimeter per second of oxygen molecules with a wall at a pressure of 1 atm and 25° C.

SOLUTION:

$$\rho = \frac{PN_A}{RT} = \frac{(101{,}325 \text{ Nm}^{-2})(6.022 \times 10^{23} \text{ mol}^{-1})}{(8.3144 \text{ JK}^{-1} \text{mol}^{-1})(298.15 \text{ K})}$$

$$\rho = 2.46 \times 10^{25} \text{ molecules m}^{-3}$$

$$Z = \rho \left(\frac{RT}{2\pi M}\right)^{1/2}$$

$$= (2.46 \times 10^{25} \text{ molecules m}^{-3}) \left[\frac{(8.3144 \text{ JK}^{-1}\text{mol}^{-1})(298.15\text{K})}{2\pi (32 \times 10^{-3} \text{ Kg mol}^{-1})}\right]^{1/2}$$

$$= 2.73 \times 10^{27} \text{ molecules m}^{-2} \text{ s}^{-1}$$

$$= 2.73 \times 10^{24} \text{ molecules cm}^{-2} \text{ s}^{-1}$$

Strictly speaking "molecules" is not a unit and need not be indicated here. However, it is sometimes convenient to indicate what we are counting.

14.9 A Knudsen cell containing crystalline benzoic acid ($M = 122$ g mol^{-1}) is carefully weighed and placed in an evacuated chamber thermostat at 70°C for 1 hr. The circular hole through which effusion occurs is 0.60 mm in diameter. Calculate the sublimation pressure of benzoic acid at 70°C in torr from the fact that the weight loss is 56.7 mg.

SOLUTION:

$$P = W \left(\frac{2\pi RT}{M}\right)^{1/2}$$

$$= \frac{56.7 \times 10^{-6} \text{ Kg}}{(60 \times 60\text{s})\pi(0.3 \times 10^{-3}\text{m})^2} \left[\frac{2\pi (8.314 \text{ JK}^{-1}\text{mol}^{-1})(343\text{K})}{122 \times 10^{-3} \text{ Kg mol}^{-1}}\right]^{1/2}$$

$$= 21.3 \text{ Nm}^{-2}$$

$$= \frac{(21.3 \text{ Nm}^{-2})(760 \text{ torr atm}^{-1})}{101{,}325 \text{ N m}^{-2} \text{ atm}^{-1}}$$

$$= 0.160 \text{ torr}$$

14.10 Large vacuum chambers have been built fo testing space vehicles at 10^{-8} torr. Calculate (a) the mean-free path of nitrogen at this pressure and (b) the number of molecular impacts per m^2 of wall per second at 25°C. $\sigma_{N_2} = 0.375$ nm.

SOLUTION:

$$n = \frac{PN_A}{RT} = \frac{(10^{-8}\ \text{torr})(6.022 \times 10^{23}\ \text{mol}^{-1})}{(760\ \text{torr atm}^{-1})(0.082\ \text{L atm K}^{-1}\text{mol}^{-1})(298\text{K})}$$

$$= 3.24 \times 10^{11}\ \text{L}^{-1}$$

$$= 3.24 \times 10^{14}\ \text{m}^{-3}$$

(a) $$l = \frac{1}{\sqrt{2}\,\pi\,\sigma^2 n}$$

$$= \frac{1}{\sqrt{2}\,\pi\,(3.75 \times 10^{-10}\ \text{m})^2 (3.24 \times 10^{14}\ \text{m}^{-3})}$$

$$= 4940\ \text{m}$$

(b) $$Z = \rho \left(\frac{RT}{2\pi M}\right)^{1/2}$$

$$= (3.24 \times 10^{14}\ \text{m}^{-3}) \left[\frac{(8.314\ \text{J K}^{-1}\text{mol}^{-1})(298\text{K})}{2\pi\ 28 \times 10^{-3}\ \text{Kg mol}^{-1}}\right]^{1/2}$$

$$= 3.84 \times 10^{16}\ \text{m}^{-2}\ \text{s}^{-1}$$

14.12 (a) Calculate the mean free path for hydrogen gas ($\sigma = 0.247$ nm) at 1 atm and 10^{-3} torr at 25°C. (b) Repeat the calculation for chlorine gas ($\sigma = 0.496$ nm).

SOLUTION:

At 1atm $\rho = \frac{N_A P}{RT} = \frac{(6.022 \times 10^{23}\ \text{mol}^{-1})(101,325\ \text{N m}^{-2})}{(8.314\ \text{J K}^{-1}\text{mol}^{-1})(298\ \text{K})}$

$= 2.46 \times 10^{25}\ \text{m}^{-3}$

At 10^{-3} torr $\rho = \frac{(6.022 \times 10^{-23}\ \text{mol}^{-1})(\frac{10^{-3}}{760}\ 101,325\ \text{N m}^{-2})}{(8.314\ \text{J K}^{-1}\ \text{mol}^{-1})(298\ \text{K})}$

$= 3.24 \times 10^{19}\ \text{m}^{-3}$

(a) $l = \frac{1}{2^{1/2}\pi\sigma^2\rho}$

$= \frac{1}{2^{1/2}\pi(0.247 \times 10^{-9}\ \text{m})^2(2.46 \times 10^{25}\ \text{m}^{-3})}$

$= 1.50 \times 10^{-7}\ \text{m}$ at 1 atm

$l = \frac{1}{2^{1/2}\pi(0.247 \times 10^{-9}\ \text{m})^2(3.24 \times 10^{19}\ \text{m}^{-3})}$

$= 0.114\ \text{m}$ at 10^{-3} torr

(b) $l = \frac{1}{2^{1/2}\pi(0.496 \times 10^{-9}\ \text{m})^2(2.46 \times 10^{25}\ \text{m}^{-3})}$

$= 3.72 \times 10^{-8}\ \text{m}$ at 1 atm

$l = \frac{1}{2^{1/2}\pi(0.496 \times 10^{-9}\ \text{m})^2(3.24 \times 10^{19}\ \text{m}^{-3})}$

$= 0.028\ \text{m}$ at 10^{-3} torr

14.13 Consider an atomic beam of potassium passing through a scattering gas of Ar contained in a cell of 1 cm length at 0°C. Assuming a collision cross section of $6 \times 10^{-18}\ m^2$ for potassium-argon collisions, calculate the pressure of argon in torr required to produce an attenuation of the beam of 25%.

SOLUTION:

$$I_A = I_A^{\circ} e^{-Q_{AB} n_B L}$$

$$n_B = \frac{\ln (I_A^{\circ}/I_A)}{Q_{AB} L}$$

$$= \frac{2.303 \log (100/75)}{(600 \times 10^{-20}\ m^2)(10^{-2} m)}$$

$$= 4.80 \times 10^{18}\ m^{-3} = 4.80 \times 10^{15}\ L^{-1}$$

$$P = \frac{n_B RT}{N_A} = \frac{(4.80 \times 10^{15} L^{-1})(0.082\ L\ atm\ K^{-1} mol^{-1})(273\ K)}{(6.022 \times 10^{23}\ mol^{-1})} \times (760\ torr\ atm^{-1})$$

$$= 1.4 \times 10^{-4}\ torr$$

14.14 Oxygen is contained in a vessel at 2 torr pressure and 25°C. Calculate (a) the number of collisions between molecules per second per cubic centimeter, (b) the mean free path. $\sigma_{O_2} = 0.361$ nm.

SOLUTION:

$$n = \frac{PN_A}{RT} = \frac{(2\ \text{torr})(6.022 \times 10^{23}\ \text{mol}^{-1})}{(760\ \text{torr atm}^{-1})(0.082\ \text{L atm K}^{-1}\text{mol}^{-1}) \times (298\ \text{K})}$$

$$= 6.48 \times 10^{19}\ \text{L}^{-1}$$

$$= 6.48 \times 10^{22}\ \text{m}^{-3}$$

$$\langle v \rangle = \left(\frac{8RT}{\pi M}\right)^{1/2}$$

$$= \left[\frac{8(8.314\ \text{J K}^{-1}\ \text{mol}^{-1})(298\text{K})}{\pi(32 \times 10^{-3}\ \text{Kg})}\right]^{1/2}$$

$$= 444\ \text{ms}^{-1}$$

(a)
$$Z_{11} = \frac{1}{2^{1/2}} \rho^2 \pi \sigma^2 \langle v \rangle$$

$$= \frac{1}{2^{1/2}} (6.48 \times 10^{22}\ \text{m}^{-3})^2 \pi (0.361 \times 10^{-9}\ \text{m})^2 (444\ \text{ms}^{-1})$$

$$= 5.40 \times 10^{29}\ \text{m}^{-3}\ \text{s}^{-1}$$

(b)
$$l = \frac{1}{\sqrt{2}\,\pi \sigma^2 \rho} = \frac{1}{\sqrt{2}\,\pi (3.61 \times 10^{-10}\ \text{m})^2 (6.48 \times 10^{22}\ \text{m}^{-3})}$$

$$= 2.67 \times 10^{-5}\ \text{m}$$

14.15 The pressure in interplanetary space is estimated to be of the order of 10^{-16} torr. Calculate (a) the average number of molecules per cubic centimeter, (b) the number of collisions per second per molecule, and (c) the mean free path in miles. Assume that only hydrogen atoms are present and that the temperature is 1000 K. Assume $\sigma = 0.2$ nm.

SOLUTION:

(a) $$\rho = \frac{N_A P}{RT} = \frac{(6.022\times10^{23}\text{ mol}^{-1})(\frac{10^{-16}}{760}\ 101,325\text{ N m}^{-2})}{(8.314\text{ J K}^{-1}\text{ mol}^{-1})(10^3\text{ K})}$$

$$= 0.965\times10^6\text{ m}^{-3}$$

$$= 0.966\text{ cm}^{-3}$$

(b) $$\langle v\rangle = \left(\frac{8RT}{\pi M}\right)^{1/2}$$

$$= \left[\frac{8(8.314\text{ J K}^{-1}\text{ mol}^{-1})(10^3\text{ K})}{\pi\ (1\times10^{-3}\text{ Kg mol}^{-1})}\right]^{1/2}$$

$$= 4600\text{ ms}^{-1}$$

$$Z_1 = 2^{1/2}\rho\pi\sigma^2\langle v\rangle$$

$$= 2^{1/2}(0.966\times10^6\text{ m}^{-3})\pi(0.2\times10^{-9}\text{m})^2(4600\text{ ms}^{-1})$$

$$= 7.90\times10^{-10}\text{ s}^{-1}$$

(c) $$l = \frac{1}{2^{1/2}\pi\sigma^2\rho}$$

$$= \frac{1}{2^{1/2}\pi(0.2\times10^{-9}\text{ m})^2(0.966\times10^6)}$$

$$= 5.82\times10^{12}\text{ m}$$

$$= \frac{(5.82\times10^{12}\text{ m})(10^2\text{ cm m}^{-1})}{(2.54\text{ cm m}^{-1})(12\text{ in ft.}^{-1})(5280\text{ ft mile}^{-1})}$$

$$= 3.62\times10^9\text{ miles}$$

14.17 For Ne parameters of the Lennard-Jones 6-12 potential are $\epsilon/k = 35.6$ K and $\sigma = 0.275$ nm. Plot V in Kcal mol^{-1} versus r and calculate the distance r_m where $dV/dr = 0$.

SOLUTION:

$$r_m = 2^{1/6}\sigma = 2^{1/6}\, 0.275 \text{ nm} = 3.09 \text{ nm}$$

$$\epsilon = (1.987\times10^{-3}\text{ Kcal K}^{-1}\text{ mol}^{-1})(35.6\text{K}) = 70.7\times10^{-3}\text{ Kcal mol}^{-1}$$

$$V = 4\epsilon\left[\left(\frac{\sigma}{r}\right)^{12} - \left(\frac{\sigma}{r}\right)^{6}\right]$$

$$= 4\,(70.7\times10^{-3}\text{ K cal mol}^{-1})\left[\left(\frac{2.75}{r}\right)^{12} - \left(\frac{2.75}{r}\right)^{6}\right]$$

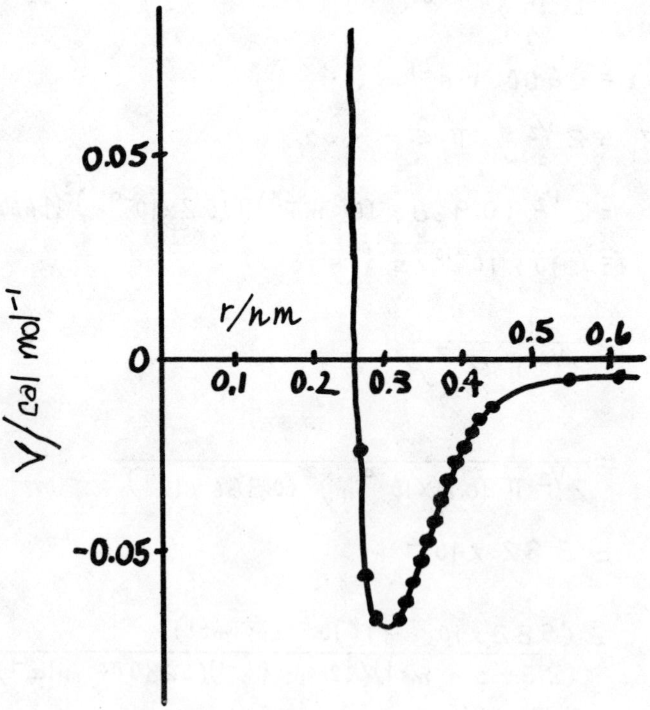

14.18 For rigid sphere molecules the coefficient of viscosity is given by equation 14.66. Given that for N_2 $\sigma = 0.375$ nm, what is the coefficient of viscosity of N_2 at 0°C ?

SOLUTION:

$$m = \frac{2(14 \times 10^{-3} \text{ kg mol}^{-1})}{6.022 \times 10^{23} \text{ mol}^{-1}} = 4.65 \times 10^{-26} \text{ kg}$$

$$\eta = \frac{5}{16} \frac{(\pi m k T)^{1/2}}{\pi \sigma^2}$$

$$= \frac{5}{16} \frac{[\pi (4.65 \times 10^{-26} \text{ kg})(1.38 \times 10^{-23} \text{ J K}^{-1})(273 \text{ K})]^{1/2}}{\pi (0.375 \times 10^{-9} \text{ m})^2}$$

$$= 1.66 \times 10^{-5} \text{ Pa s}$$

14.19 3.87781×10^{-2} eV

14.20 28 torr

14.21 0.904 atm

14.22 0.1991

14.23 481, 445 and 394 m s^{-1}

14.24 418, 400, and 440 m s^{-1}

14.25 7.23×10^4 cm s^{-1} 671K

14.26 (a) 1.02×10^5 cm s^{-1} (b) 3.52×10^4 cm s^{-1}

14.27 2780 s

14.28 (a) 1.136×10^{22} molecules cm^{-2} s^{-1}

(b) 0.0629 g cm^{-2} min^{-1}

14.29 7.26 kcal mol^{-1}

14.30 0.0713 g

14.31 (a) 3.54×10^{6} cm^{-3} (b) 4.90×10^{7} cm

14.32 (a) 28 mm^2 (b) 14 nm^2

(c) The <u>cis</u> compound has a large dipole moment while the <u>trans</u> compound has no permanent dipole moment. Since CsCl is a dipole molecule, it interacts more strongly with the polar isomer of dichloroethylene.

14.33 4.86 torr

14.34 9.7×10^{-11} s

14.35 (a) 7.31×10^{9} s^{-1} (b) 8.97×10^{28} cm^{-3} s^{-1}

(c) 0.354 (d) 4

14.36 5.80×10^{28}

14.37

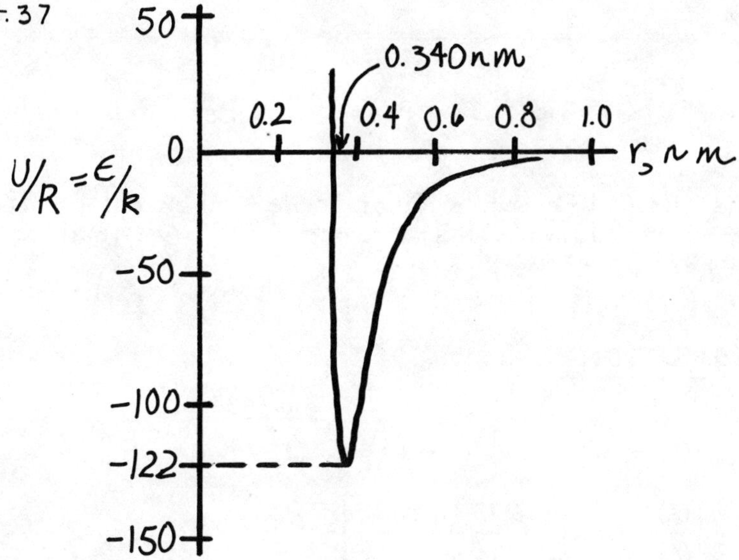

14.38 0.429 nm

14.39 (a) 0.217 nm (b) 1.27×10^{-4} m^2 s^{-1}

14.40 1.302 Pa s

CHAPTER 15 Kinetics: Gas Phase and Basic Relationships

15.1 The half life of a first-order chemical reaction $A \rightarrow B$ is 10 min. What percent of A remains after 1 hr?

SOLUTION:

$$(A) = (A_o)\, e^{-kT} = (A)_o\, e^{-\frac{0.693t}{t_{1/2}}}$$

$$\frac{(A)}{(A)_o} = \exp\left[\frac{-0.693\,(60\text{ m})}{10\text{ m}}\right] = 0.0156$$

Thus 1.56% remains after one hour.

15.2 The following data were obtained on the rate of hydrolysis of 17% sucrose in 0.099 mol L^{-1}. HCl aqueous solution at 35°C.

t/min	9.82	59.60	93.18	142.9	294.8	589.4
Sucrose remaing %	96.5	80.3	71.0	59.1	32.8	11.1

SOLUTION:

$$\log \frac{(A)}{(A)_o} = \frac{-kT}{2.303}$$

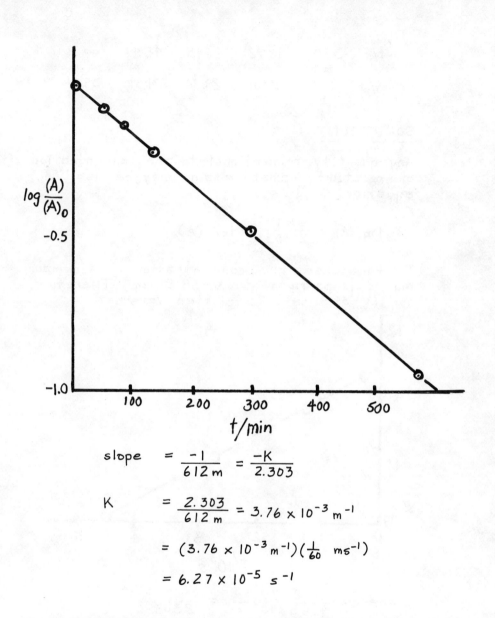

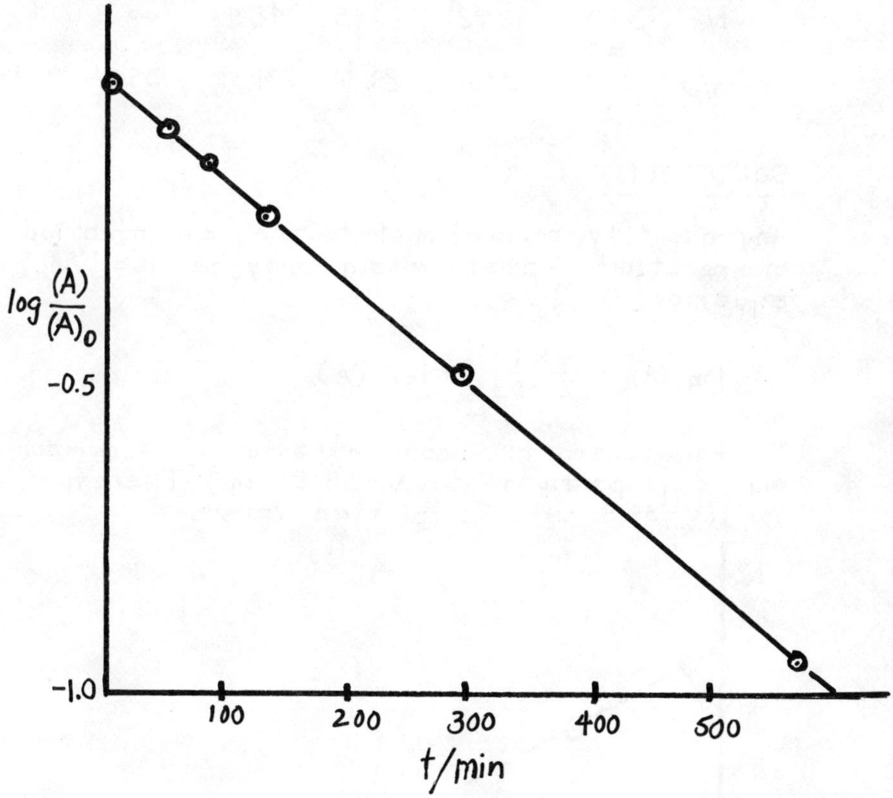

$$\text{slope} = \frac{-1}{612\ m} = \frac{-K}{2.303}$$

$$K = \frac{2.303}{612\ m} = 3.76 \times 10^{-3}\ m^{-1}$$

$$= (3.76 \times 10^{-3}\ m^{-1})(\tfrac{1}{60}\ ms^{-1})$$

$$= 6.27 \times 10^{-5}\ s^{-1}$$

15.3 Methyl acetate is hydrolyzed in approximately 1 mol L^{-1} HCl at 25° C. Aliquots of equal volume are removed at intervals and titrated with a solution of NaOH. Calculate the first order rate constant from the following experimental data:

t/s	339	1242	2745	4546	∞
V/cm^3	26.34	27.80	29.70	31.81	39.81

SOLUTION:

Any quantity proportional to the concentration of reactant A that remains may be used in equation

$$\log (A) = \frac{-KT}{2.303} + \log (A)_0$$

In this case the concentration of A remaining is proportional to $V - 39.81\ cm^3$. Therefore $\log (V - 39.81\ cm^3)$ is plotted versus t.

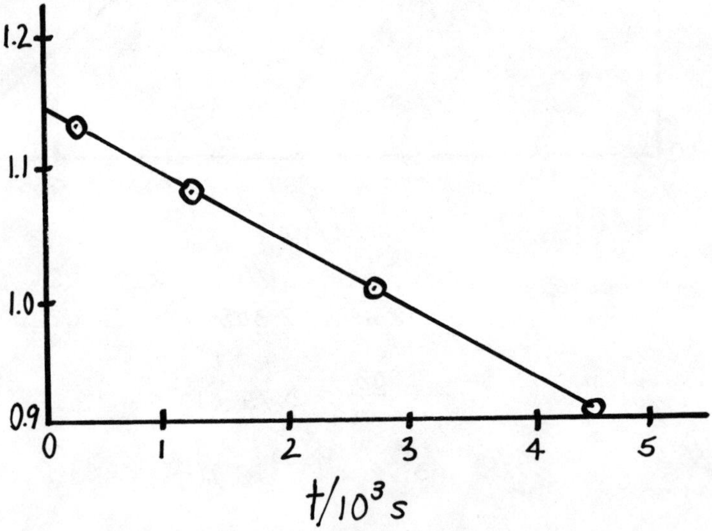

$$\text{slope} = \frac{-0.247}{4.65 \times 10^3\ s} = \frac{-k}{2.303}$$

$$k = \frac{(2.303)(0.247)}{4.65 \times 10^3\ s} = 1.22 \times 10^{-4}\ s^{-1}$$

15.4 Prove that in a first-order reaction, where $dn/dt = -kn$, the average life, that is, the average life expectancy of the molecules, is equal to 1/k.

SOLUTION:

$$\text{Average life} = \frac{\int_0^\infty n\,dt}{n_0}$$

$$= \frac{-\frac{1}{k}\int_0^\infty \frac{dn}{dt}\,dt}{n_0}$$

$$= \frac{-\frac{1}{k}(0-n_0)}{n_0}$$

$$= \frac{1}{k}$$

15.5 It is found that the decomposition of HI to $H_2 + I_2$ at 508°C has a half-life of 135 min when the initial pressure of HI is 0.1 atm and 13.5 min when the pressure is 1 atm. (a) Show that this proves that the reaction is second order. (b) What is the value of the rate constant in L mol^{-1} s^{-1}? (c) What is the value of the rate constant in atm^{-1} s^{-1}?

SOLUTION:

(a) For a second order reaction $t_{1/2} = 1/k(A)_0$ where $(A)_0$ is the initial concentration or pressure of the reactant. This is in agreement with the fact that the half life is reduced by a factor of 10 when the pressure is increased by a factor of 10.

(b) $c = P/RT = (1\text{ atm})/(0.08205\text{ L atm K}^{-1}\text{ mol}^{-1})(781.15\text{ K})$

$= 1.56 \times 10^{-2}$ mol L^{-1}

$$k = 1/at_{\frac{1}{2}} = 1/(1.56 \times 10^{-2}\ mol\ L^{-1})(13.5 \times 60\ s)$$

$$= 7.91 \times 10^{-2}\ L\ mol^{-1}\ s^{-1}$$

(c) $k = 1/at_{\frac{1}{2}} = 1/(1\,atm)(13.5 \times 60 s) = 1.24 \times 10^{-3}\ atm^{-1}\ s^{-1}$

15.6 The reaction between propionaldehyde and hydrocyanic acid has been studied at 25°C by W. J. Svirbely and J. F. Roth [J. Am. Chem. Soc., 75, 3106 (1953)]. In a certain aqueous solution at 25°C the concentrations at various times were as follows:

t/min	2.78	5.33	8.17	15.23	19.80	∞
(HCN)/mol L^{-1}	0.0990	0.0906	0.0830	0.0706	0.0653	0.0424
(C_3H_7CHO)/mol L^{-1}	0.0566	0.0482	0.0406	0.0282	0.0229	0.0000

What is the order of the reaction and the value of the rate constant K?

SOLUTION:

The data does not give a linear log versus t plot, and so the data are tested in the integrated equation for a second order reaction. Equation 15.20 is

$$\frac{1}{[(A)_0 - (B)_0]} \ln \frac{(A)(B)_0}{(A)_0(B)} = kT$$

In order to get $(A)_0$ and $(B)_0$ for t = 0, the clock is started at the first experimental point. This yields the following data to be plotted.

t/ min	0	2.55	5.39	12.45	17.02
$\frac{1}{[(A)_0-(B)_0]}\ln\frac{(A)(B)_0}{(A)_0(B)}$	0	1.695	3.677	8.458	11.523

The slope of this plot, 0.675 L mol^{-1} min^{-1}, is the second order rate constant.

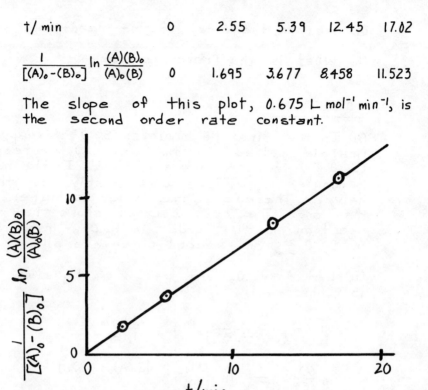

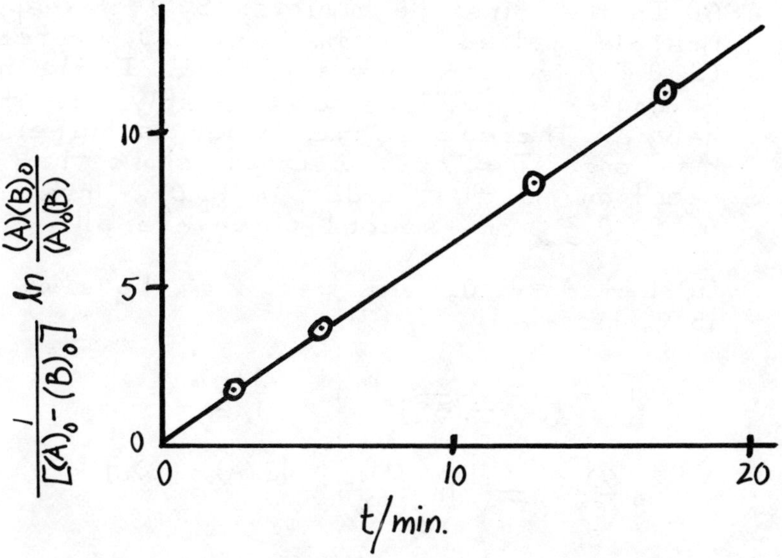

15.8 Hydrogen peroxide reacts with thiosulfate ion in slightly acid solution as follows:

$$H_2O_2 + 2S_2O_3^{2-} + 2H^+ \rightarrow 2H_2O + S_4O_6^{2-}$$

This reaction rate is independent of the hydrogen-ion concentration in the pH range 4-6. The following data were obtained at 25 °C and pH 5.0:

Initial concentrations: $(H_2O_2) = 0.03680$ mol L^{-1} ; $(S_2O_3^{2-}) = 0.02040$ mol L^{-1}

t/ min	16	36	43	52
$(S_2O_3^{2-})/10^{-3}$ mol L^{-1}	10.30	5.18	4.16	3.13

(a) What is the order of the reaction?

(b) What is the rate constant?

SOLUTION:

(a) In the first 16 minutes, $(S_2O_3^{2-})$ is approximately halved. In the next 20 minutes, $(S_2O_3^{2-})$ is approximately halved. In the next 16 minutes $(S_2O_3^{2-})$ is considerably less than halved. Therefore, the order is higher than one. The next section shows that the reaction is first order in H_2O_2, first order in $S_2O_3^{2-}$, and second order overall.

(b) Let $A = H_2O_2$ and $B = S_2O_3^{2-}$. Equation 15.19 becomes

$$kT = \frac{1}{[2(A)_0 - (B)_0]} \ln \frac{(A)(B)_0}{(A)_0(B)}$$

$$\ln \frac{(A)}{(B)} = \ln \frac{(A)_0}{(B)_0} + [2(A)_0 - (B)_0]kT$$

t/min	0	16	36	43	52
$(B)/10^{-3}$ mol L^{-1}	20.40	10.30	5.18	4.16	3.13
$(A)/10^{-3}$ mol L^{-1}	36.80	31.75	29.19	28.68	28.17
$\ln \frac{(A)}{(B)}$	0.590	1.126	1.729	1.931	2.197

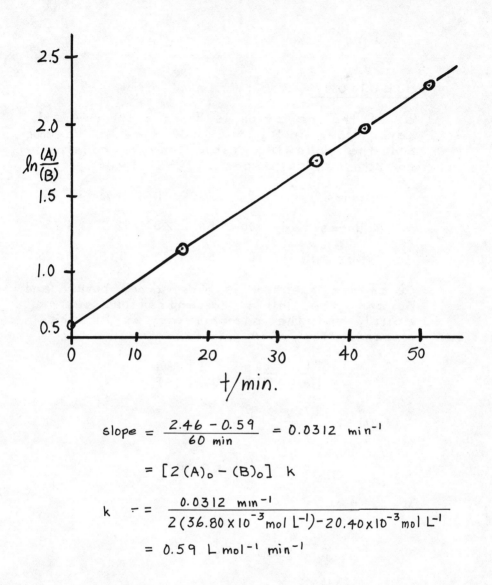

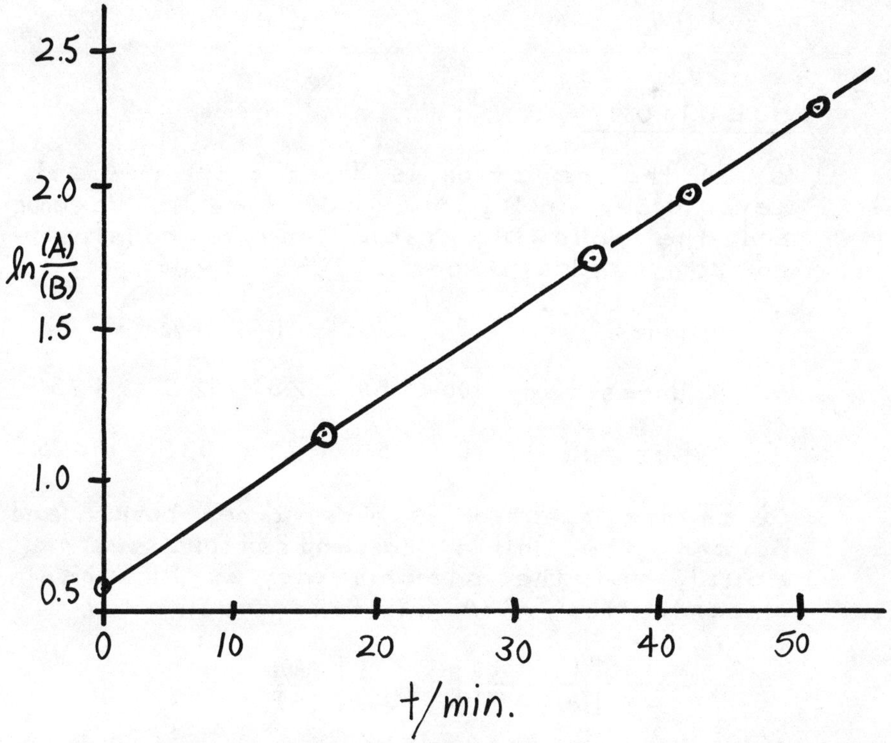

$$\text{slope} = \frac{2.46 - 0.59}{60\ \text{min}} = 0.0312\ \text{min}^{-1}$$

$$= [2(A)_0 - (B)_0]\ k$$

$$k = \frac{0.0312\ \text{min}^{-1}}{2(36.80 \times 10^{-3}\ \text{mol L}^{-1}) - 20.40 \times 10^{-3}\ \text{mol L}^{-1}}$$

$$= 0.59\ \text{L mol}^{-1}\ \text{min}^{-1}$$

15.9 A solution of A is mixed with an equal volume of a solution of B containing the same number of moles, and the reaction A + B = C occurs. At the end of 1 hr A is 75% reacted. How much of A will be left unreacted at the end of 2 hr if the reaction is (a) first order in A and zero order in B; (b) first order in both A and B; (c) zero order in both

A and B?

SOLUTION:

(a) If the reaction is first order in A and zero order in B, the half time is 1/2 hour, and the following table may be constructed doing calculations in your head.

t/hours	0	1/2	1	1 1/2	2
% Unreacted	100	50	25	12.5	6.25
% Reacted	0	50	75	87.5	93.75

(b) If the reaction is first order both A and B, and the initial concentrations are equal, and the stoichiometry is 1:1, the concentration of A will follow eqn. 15.17

$$k = \frac{1}{t}\left[\frac{1}{(A)} - \frac{1}{(A)_0}\right] = \frac{1}{t(A)_0}\left[\frac{(A)_0}{(A)} - 1\right]$$

$$k(A)_0 = \frac{1}{t}\left[\frac{(A)_0}{(A)} - 1\right]$$

$$= \frac{1}{1\,hr}\left[\frac{100}{25} - 1\right] = 3\ hr^{-1}$$

After 2 hr

$$3\ hr^{-1} = \frac{1}{2\,hr}\left[\frac{100}{(A)} - 0\right]$$

$$(A) = \frac{100}{7} = 14.3\ \%$$

(c) If the reaction is zero order in both A and B, both will be completely gone in 1 1/3 hr.

15.11 For the reaction $2A = B + C$ the rate law for the forward reaction is

$$-\frac{d(A)}{dt} = k(A)$$

Give two possible rate laws for the reverse reaction.

SOLUTION:

$$K = \frac{(B)(C)}{(A)^2}$$

At equilibrium $-\frac{d(A)}{dt} = 0 = k_f(A)^2 - k_r(B)(C)$

$$0 = k_f(A) - k_r(B)(C)/(A)$$

Possible rate law for the reverse reaction
$k_r(B)(C)/(A)$

$$K = \frac{(B)^{1/2}(C)^{1/2}}{(A)}$$

At equilibrium $-\frac{d(A)}{dt} = 0 = k_f(A) - k_r(B)^{1/2}(C)^{1/2}$

Possible rate law for the reverse reaction
$k_r(B)^{1/2}(C)^{1/2}$

$$K = \frac{(B)^2(C)^2}{(A)^4}$$

At equilibrium $\frac{-d(A)}{dt} = 0 = k_f(A)^4 - k_r(B)^2(C)^2$

$$0 = k_f(A) - k_r\frac{(B)^2(C)^2}{(A)^3}$$

Possible rate law for the reverse reaction
$k_r(B)^2(C)^2/(A)^3$

15.12 The following table gives kinetic data [Y. T. Chia and R.E. Connick, J. Phys. Chem., 63, 1518 (1959)] for the following reaction at 25°C.

$$OCl^- + I^- = OI^- + Cl^-$$

(OCl^-)	(I^-) mol L^{-1}	(OH^-)	$\frac{d(IO^-)}{dt}/10^{-4}$ mol $L^{-1}s^{-1}$
0.0017	0.0017	1.00	1.75
0.0034	0.0017	1.00	3.50
0.0017	0.0034	1.00	3.50
0.0017	0.0017	0.5	3.50

What is the rate law for the reaction and what is the value of the rate constant?

SOLUTION:

When other concentrations are held constant, doubling (OCl^-) doubles the rate, doubling (I^-) doubles the rate, and halving (OH^-) doubles the rate. Therefore the rate law is

$$\frac{d(OI^-)}{dt} = \frac{k(OCl^-)(I^-)}{(OH^-)}$$

Substituting the values for the first experiment

$$1.75 \times 10^{-4} \text{ mol L}^{-1} \text{ s}^{-1} = \frac{k(0.0017 \text{ mol L}^{-1})(0.0017 \text{ mol L}^{-1})}{(1 \text{ mol L}^{-1})}$$

$$k = 61 \text{ s}^{-1}$$

15.13 When an optically active substance is isomerized, the optical rotation decreases from that of the original isomer to zero in a first-order manner. In a given case the half-time for this process is found to be 10 min.

Calculate the rate constant for the conversion of one isomer to another.

SOLUTION:

$$t_{1/2} = \frac{0.693}{k_1 + k_{-1}} = 600 \text{ s}$$

Since $K = 1$, $k_1 = k_{-1}$

$$t_{1/2} = \frac{0.693}{2k_1} = 600 \text{ s}$$

$$k_1 = \frac{0.693}{2(600\text{s})} = 5.78 \times 10^{-4} \text{ s}^{-1}$$

15.15 The reaction $2NO + O_2 \longrightarrow 2NO_2$ is third order. Assuming that a small amount of NO_3 exists in rapid reversible equilibrium in NO and O_2 and that the rate-determining step is the slow bimolecular reaction $NO_3 + NO \rightarrow 2NO_2$, derive the rate equation for this mechanism.

SOLUTION:

$$NO + O_2 \rightleftharpoons NO_3 \qquad K_{eq} = \frac{(NO_3)}{(NO)(O_2)}$$

$$NO_3 + NO \xrightarrow{k_2} 2NO_2 \quad \text{(rate determing)}$$

$$\frac{d(NO_2)}{dt} = 2k_2 (NO_3)(NO)$$

$$= 2k_2 K_{eq} (NO)^2 (O_2)$$

$$= k' (NO)^2 (O_2)$$

15.16 Set up the rate expression for the following mechanism:

$$A \underset{k_2}{\overset{k_1}{\rightleftharpoons}} B$$

$$B + C \xrightarrow{k_3} D$$

If the concentration of B is small compared with the concentrations of A, C, and D, the steady state approximation may be used to derive the rate law. Show that this reaction may follow the first-order equation at high pressures and the second-order equations at low pressures.

SOLUTION:

Since the concentration of B is small, it is assumed to be in a steady state.

$$\frac{d(B)}{dt} = k_1(A) - [k_2 + k_3(C)](B) = 0$$

$$(B) = \frac{k_1(A)}{k_2 + k_3(C)}$$

$$\frac{d(D)}{dt} = k_3(B)(C) = \frac{k_1 k_3 (A)(C)}{k_2 + k_3(C)}$$

At high pressures,

$$k_3(C) \gg k_2, \quad \frac{d(D)}{dt} = k_1(A)$$

At low pressures,

$$k_2 \gg k_3(C), \quad \frac{d(D)}{dt} = \frac{k_1 k_3}{k_2}(A)(C)$$

15.17 The reaction $NO_2Cl = NO_2 + \frac{1}{2}Cl_2$ is first order and appears to follow the mechanism.

$$NO_2Cl \xrightarrow{k_1} NO_2 + Cl$$

$$NO_2Cl + Cl \xrightarrow{k_2} NO_2 + Cl_2$$

(a) Assuming a steady state for the chlorine atom concentration, show that the empirical first-order rate constant can be identified with $2k_1$. (b) The following data were obtained by H. F. Cordes and H. S. Johnston [J. Am. Chem. Soc., 76, 4264 (1954)] at 180°C. In a single experiment the reaction is first order, and the empirical rate constant is represented by k. Show that the reaction is second order at these low gas pressures and calculate the second-order rate constant.

$c/10^{-8}$ mol cm^{-3}	5	10	15	20
$k/10^{-4}$ s^{-1}	1.7	3.4	5.2	6.9

SOLUTION:

(a) $$\frac{d(Cl)}{dt} = k_1(NO_2Cl) - k_2(NO_2Cl)(Cl) = 0$$

Therefore $(Cl) = \frac{k_1}{k_2}$

$$\frac{-d(NO_2Cl)}{dt} = k_1(NO_2Cl) + k_2(NO_2Cl)(Cl)$$

$$= k_1(NO_2Cl) + k_2(NO_2Cl)k_1/k_2$$

$$= 2k_1(NO_2Cl)$$

(b) At low pressures the rate determing step is the first step, and the reaction becomes second order.

$$NO_2Cl + NO_2Cl \xrightarrow{k_1'} NO_2Cl + NO_2 + Cl$$

$$\frac{-d(NO_2Cl)}{dt} = k_1'(NO_2Cl)^2$$

assuming (NO_2Cl) is approximately constant

$$k_{obs} = k_1'(NO_2Cl)$$

$$1.7 \times 10^{-4} = k_1'(5 \times 10^{-8})$$

$$k_1' = 3.4 \times 10^3 \text{ cm}^3 \text{ mol}^{-1} \text{ s}^{-1}$$

15.18 Suppose the transformation of A to B occurs by both a reversible first-order reaction and a reversible second order reaction involving hydrogen ion:

$$A \underset{k_2}{\overset{k_1}{\rightleftharpoons}} B$$

$$A + H^+ \underset{k_4}{\overset{k_3}{\rightleftharpoons}} B + H^+$$

What is the relationship between these four rate constants?

SOLUTION:

$$\frac{(B)_{eq}}{(A)_{eq}} = \frac{k_1}{k_2}$$

$$\frac{(B)_{eq}(H^+)_{eq}}{(A)_{eq}(H^+)_{eq}} = \frac{k_3}{k_4} \quad \text{or} \quad \frac{(B)_{eq}}{(A)_{eq}} = \frac{k_3}{k_4}$$

Therefore $\frac{k_1}{k_2} = \frac{k_3}{k_4}$ or $k_1 k_4 = k_2 k_3$

15.20 Isopropenyl allyl ether in the vapor state isomerizes to allyl acetone according to a first-order rate equation. The following equation gives the influence of temperature on the rate constant (in s^{-1}):

$$k = 5.4 \times 10^{11} e^{-29,300/RT}$$

where the activation energy is expressed in cal mol^{-1}. At 150° C how long will it take to build up a partial pressure of 300 torr of allyl acetone starting with 760 torr isopropenyl allyl ether [L. Stein and G. W Murphy, J. Am. Chem. Soc., 74, 1041 (1952)]?

SOLUTION

$$k = (5.4 \times 10^{11}\ s^{-1}) e^{-\frac{29,300}{(1.987)(423.15)}}$$

$$= 3.96 \times 10^{-4}\ s^{-1}$$

$$t = \frac{1}{k} \ln \frac{P^\circ}{P}$$

$$= \frac{1}{3.96 \times 10^{-4}\ s^{-1}} \ln \frac{760\ \text{torr}}{460\ \text{torr}}$$

$$= 1270\ s$$

15.21 The hydrolysis of $(CH_2)_6\, C\genfrac{}{}{0pt}{}{Cl}{CH_3}$ in 80% ethanol follows the first-order rate equation. The values of the specific reaction-rate constants, as determined by H. C. Brown and M. Borkowski [J. Am. Chem. Soc., 74, 1896 (1952)], are as follows:

$t/°C$	0	25	35	45
k/s^{-1}	1.06×10^{-5}	3.19×10^{-4}	9.86×10^{-4}	2.92×10^{-3}

(a) Plot log k against 1/T ; (b) calculate the activation energy; (c) calculate the preexponential factor.

SOLUTION:

(a)

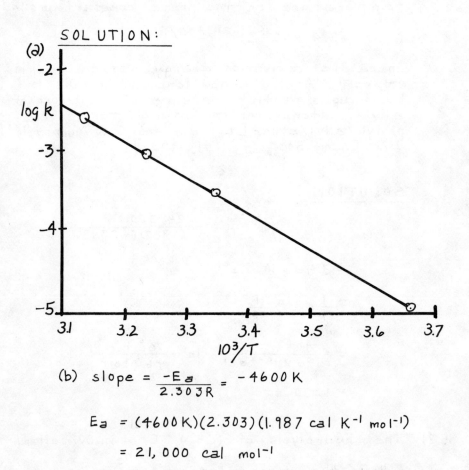

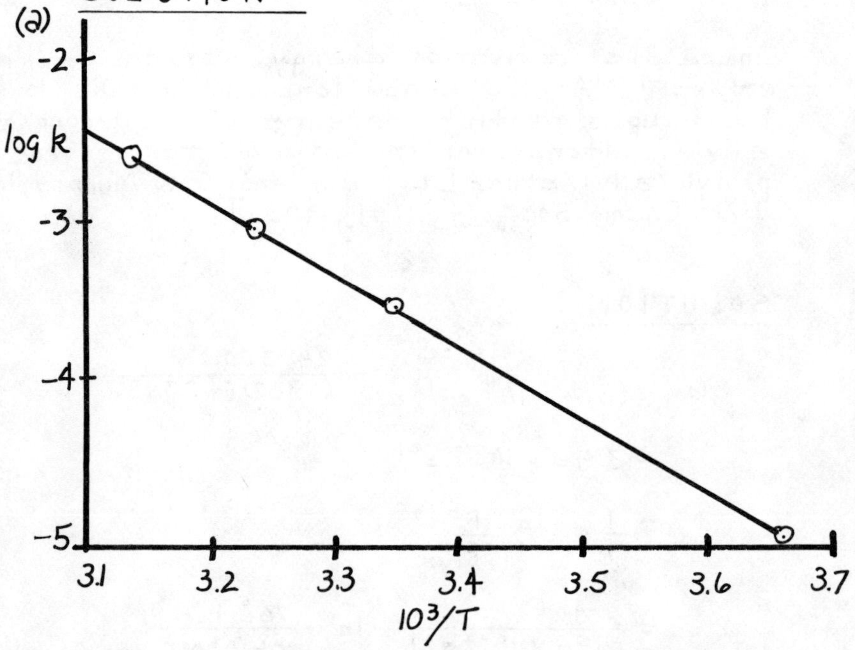

(b) $\text{slope} = \dfrac{-E_a}{2.303R} = -4600\text{ K}$

$$E_a = (4600\text{ K})(2.303)(1.987\text{ cal K}^{-1}\text{ mol}^{-1})$$

$$= 21,000\text{ cal mol}^{-1}$$

(c) Taking the 35 °C point and $E_a = 21,000$ cal mol^{-1}

$$\log k = \frac{-E_a}{2.303 RT} + \log A$$

$$-3.006 = \frac{-21,000 \text{ cal mol}^{-1}}{2.303 (1.987 \text{ cal K}^{-1} \text{ mol}^{-1})(308.15 \text{K})} + \log A$$

$$A = 7.8 \times 10^{11} \text{ s}^{-1}$$

15.23 The pre-exponential factor for the termolecular reaction

$$2NO + O_2 \longrightarrow 2NO_2$$

is 10^9 cm^6 mol^{-2} s^{-1}. What is the value in L^2 mol^{-2} s^{-1}?

SOLUTION:

$$(10^9 \text{ cm}^6 \text{ mol}^{-2} \text{ s}^{-1}) (10^{-1} \text{ dm cm}^{-1})^6$$

$$= 10^3 \text{ dm}^6 \text{ mol}^{-2} \text{ s}^{-1}$$

$$= 10^3 \text{ L}^2 \text{ mol}^{-2} \text{ s}^{-1}$$

15.24 For the two parallel reactions $A \xrightarrow{k_1} B$ and $A \xrightarrow{k_2} C$ show that the activation energy E' for the disappearance of A is given in terms of the activation energies E_1 and E_2 for the two paths by

$$E' = \frac{k_1 E_1 + k_2 E_2}{k_1 + k_2}$$

SOLUTION:

The rate equation for A is

$$\frac{-d(A)}{dt} = k_1(A) + k_2(A) = (k_1 + k_2)(A) = k'(A)$$

where $k' = k_1 + k_2 = A e^{-E'/RT}$

$$\frac{d \ln k'}{dT} = \frac{E'}{RT^2} = \frac{d \ln(k_1+k_2)}{dT} = \frac{d(k_1+k_2)}{(k_1+k_2)dT}$$

$$= \frac{1}{(k_1+k_2)}\left(\frac{dk_1}{dT} + \frac{dk_2}{dT}\right)$$

$$= \frac{1}{(k_1+k_2)}\left[k_1 \frac{d \ln k_1}{dT} + k_2 \frac{d \ln k_2}{dT}\right]$$

$$= \frac{1}{(k_1+k_2)}\left[\frac{k_1 E_1}{RT^2} + \frac{k_2 E_2}{RT^2}\right]$$

$$E' = \frac{k_1 E_1 + k_2 E_2}{k_1 + k_2}$$

15.25 For the mechanism

$$A + B \underset{k_2}{\overset{k_1}{\rightleftharpoons}} C$$

$$C \xrightarrow{k_3} D$$

(a) Derive the rate law using the steady state approximation to eliminate the concentration of C. (b) Assuming that $k_3 \ll k_2$, express the preexponential factor A and E_a for the apparent second order rate constant in terms of A_1, A_2, and A_3 and E_{a1}, E_{a2} and E_{a3} for the three steps.

SOLUTION:

(a) $$\frac{d(C)}{dt} = k_1 (A)(B) - (k_2 + k_3)(C) = 0$$

$$\frac{d(D)}{dt} = k_3 (C) = \frac{k_1 k_3 (A)(B)}{k_2 + k_3}$$

(b) For $k_2 \gg k_3$,

$$k_{app} = \frac{k_1 k_3}{k_2} = \frac{A_1 e^{-E_{a1}/RT} A_3 e^{-E_{a3}/RT}}{A_2 e^{-E_{a2}/RT}}$$

$$= \frac{A_1 A_3}{A_2} e^{-(E_{a1} + E_{a3} - E_{a2})/RT}$$

$$A_{app} = \frac{A_1 A_3}{A_2}$$

$$E_{app} = E_{a1} + E_{a3} - E_{a2}$$

15.26 The thermal decomposition of gaseous acetaldehyde is a second-order reaction. The value of E_a is 45,500 cal mol^{-1}, and the molecular diameter of the acetaldehyde molecule is 5×10^{-8} cm. (a) Calculate the number of molecules colliding per cm^3 per second at 800 K and 760 torr pressure. (b) Calculate k in L mol^{-1} s^{-1}.

SOLUTION:

(a) ρ = number of molecules per cubic meter

$$= \frac{PN_A}{RT}$$

$$= \frac{(1\,\text{atm})(6.022 \times 10^{23}\,\text{mol}^{-1})(10^3\,\text{L m}^{-3})}{(0.08205\,\text{L atm K}^{-1}\,\text{mol}^{-1})(800\,\text{K})}$$

$$= 9.17 \times 10^{24}\,\text{m}^{-3}$$

$$Z_{11} = \frac{1}{2^{1/2}} \rho^2 \pi \sigma^2 \langle v \rangle$$

$$= 2 \rho^2 \sigma^2 \left(\frac{\pi k T}{m}\right)^{1/2}$$

$$= 2 \rho^2 \sigma^2 \left(\frac{\pi RT}{M}\right)^{1/2}$$

$$= 2\,(9.17 \times 10^{24}\,\text{m}^{-3})(5 \times 10^{-10}\,\text{m})^2 \left[\frac{\pi(8.314\,\text{J K}^{-1}\text{mol}^{-1})(800\,\text{K})}{44.05 \times 10^{-3}\,\text{kg mol}^{-1}}\right]^{1/2}$$

$$= 2.9 \times 10^{34}\,\text{m}^{-3}\,\text{s}^{-1}$$

$$= (2.9 \times 10^{34}\,\text{m}^{-3}\,\text{s}^{-1})(10^{-6}\,\text{m}^3\,\text{cm}^{-3})$$

$$= 2.9 \times 10^{28}\,\text{cm}^{-3}\,\text{s}^{-1}$$

(b) $k = \dfrac{(10^3 \text{ L m}^{-3}) N_A Z_{11} e^{-E_0/RT}}{\rho^2}$

$$= \frac{(10^3 \text{ L m}^{-3})(6.022 \times 10^{23} \text{ mol}^{-1})(2.9 \times 10^{34} \text{ m}^{-3} \text{ s}^{-1})}{(9.17 \times 10^{24} \text{ m}^{-3})} \times e^{\frac{-45,500}{(1.987)(800)}}$$

$$= 0.077 \text{ L mol}^{-1} \text{ s}^{-1}$$

15.27 Show that the activated complex theory yields the simple collision theory results when it is applied to the reaction of two rigid spherical molecules.

SOLUTION:

Assuming that the molecules react on their first collision (that is, the activation energy is zero), activated complex theory yields

$$k = \frac{RT}{h} \frac{q''_{AB}{}^{\ddagger}}{q_A' q_B'} \quad (1)$$

where q_A' and q_B' are molecular partition functions without the volume factor, and $q''_{AB}{}^{\ddagger}$ is the molecular partition function for the activated complex without the volume factor and without the vibrational factor.

$$q_A' = \left(\frac{2\pi m_A kT}{h^2}\right)^{3/2} \quad (2)$$

$$q_B' = \left(\frac{2\pi m_B kT}{h^2}\right)^{3/2} \quad (3)$$

$$q_{AB}'' = \left[\frac{2\pi (m_A + m_B) kT}{h^2}\right]^{3/2} \frac{8\pi^2 \mu (R_A + R_B)^2 kT}{h^2} \quad (4)$$

$$\mu = \frac{m_A m_B}{m_A + m_B} \qquad (5)$$

Substituting equations 2, 3, 4 and 5 in equation 1 yields

$$k = N_A \left(\frac{8\pi kT}{\mu}\right)^{1/2} (R_A + R_B)^2$$

which may be compared with equation 15.77

$$k = \pi \sigma^2 \left(\frac{8RT}{\pi \mu N_A}\right)^{1/2}$$

$$= \left(\frac{8\pi kT}{\mu}\right)^{1/2} (R_A + R_B)^2$$

The difference between these expressions of a fractor of N_A is simply a matter of units.

15.29 The vapor-phase decomposition of di-t-butyl peroxide is first order in the range 110-280 °C. and follows the equation

$$k = 3.2 \times 10^{16} e^{-39.100/RT}$$

where k is in s^{-1} and E_a is in cal mol^{-1}. Calculate (a) $\Delta H^{\ddagger\circ}$ and (b) $\Delta S^{\ddagger\circ}$.

SOLUTION:

(a) $\Delta H^{\ddagger\circ} = E_a - RT = 39{,}100 \text{ cal mol}^{-1} - (1.987 \text{ cal K}^{-1} \text{ mol}^{-1})(448\text{K})$

$$= 38.2 \text{ kcal mol}^{-1}$$

(b) $\Delta S^{\ddagger\circ} = R\left[\ln \frac{N_A\, h\, A}{eRT}\right]$

$$= (1.987\text{ cal K}^{-1}\text{mol}^{-1})\left[\ln \frac{(6.02\times10^{23}\text{ mol}^{-1})(6.626\times10^{-34}\text{ Js})}{(2.718)(8.314\text{ J K}^{-1}\text{mol}^{-1})(448\text{ K})} \times (3.2\times10^{16})\right]$$

$$= 14.2\text{ cal K}^{-1}\text{ mol}^{-1}$$

15.30 The apparent activation energy for the recombination of iodine atoms in argon is -1.4 kcal mol^{-1}. This negative temperature coefficient may result from the following mechanism

$$I + M = IM \qquad K = (IM)/(I)(M)$$

$$IM + I \underset{k_{-1}}{\overset{k_1}{\rightleftharpoons}} I_2 + M$$

Assuming that the first step remains at equilibrium, derive the rate equation which includes both the forward and reverse reactions. Show that the reverse reaction is bimolecular and the equilibrium constant expression for the dissociation of iodine is independent of the concentration of the third body

SOLUTION:

$$\frac{d(I_2)}{dt} = k_1 (I)(IM) - k_{-1} (I_2)(M)$$

Since $(IM) = K(I)(M)$

$$\frac{d(I_2)}{dt} = k_1 K (I)^2 (M) - k_1 (I_2)(M)$$

Thus the rate law for the reverse reaction is $k_1(I_2)(M)$. At equilibrium $d(I_2)/dt = 0$ and

$$\frac{(I_2)}{(I)^2} = \frac{k_1}{k_{-1}}\ K$$

15.32 For the mechanism

$$H_2 + X_2 \underset{k_{-1}}{\overset{k_1}{\rightleftharpoons}} 2HX$$

$$X_2 \underset{k_{-2}}{\overset{k_2}{\rightleftharpoons}} 2X$$

$$X + H_2 \underset{k_{-3}}{\overset{k_3}{\rightleftharpoons}} HX + H$$

$$H + X_2 \underset{k_{-4}}{\overset{k_4}{\rightleftharpoons}} HX + X$$

show that the steady state rate law is

$$\frac{d(HX)}{dt} = 2k_1(H_2)(X_2)\left[1 - \frac{(HX)^2}{K(H_2)(X_2)}\right]\left[1 + \frac{\dfrac{k_3}{k_1}\sqrt{\dfrac{2k_2}{k_{-2}(X_2)}}}{1 + \dfrac{k_{-3}(HX)}{k_4(X_2)}}\right]$$

SOLUTION:

There are five species, but there are two conservation equations

$$(H) + (HX) + 2(H_2) = \text{const.}$$

$$(X) + (HX) + 2(X_2) = \text{const.}$$

Therefore there are only three independent rate equations

$$\frac{d(HX)}{dt} = 2k_1(H_2)(X_2) + k_3(X)(H_2) + k_4(H)(X_2) - k_{-1}(HX)^2$$

$$-k_{-3}(HX)(H) - k_{-4}(HX)(X) \qquad (1)$$

$$\frac{d(X)}{dt} = 2k_2(X_2) + k_{-3}(HX)(H) + k_4(H)(X_2)$$

$$-k_{-2}(X)^2 - k_3(X)(H_2) - k_{-4}(HX)(X) = 0 \qquad (2)$$

$$\frac{d(H)}{dt} = k_3(X)(H_2) + k_{-4}(HX)(X) - k_{-3}(HX)(H)$$

$$-k_4(H)(X_2) = 0 \qquad (3)$$

Adding equations 2 and 3

$$(X) = \sqrt{\frac{2k_2}{k_{-2}}(X_2)}$$

Substituting in equation 3 yields

$$(H) = \frac{k_3(H_2) + k_{-4}(HX)}{k_{-3}(HX) + k_4(X_2)}\sqrt{\frac{2k_2(X_2)}{k_{-2}}}$$

Substituting in equation 1

$$\frac{d(HX)}{dt} = 2k_1(H_2)(X_2) - k_{-1}(HX)^2\left[k_4(X_2) - k_3(HX)\right]$$

$$\times \frac{\left[k_3(H_2) + k_{-4}(HX)\right]\sqrt{\frac{2k_2(X_2)}{k_{-2}}}}{k_{-3}(HX) + k_4(X_2)}$$

$$+\left[k_3(H_2) - k_{-4}(HX)\right]\sqrt{\frac{2k_2(X_2)}{k_{-2}}}\ \frac{\left[k_{-3}(HX) + k_4(X_2)\right]}{\left[k_3(HX) + k_4(X_2)\right]}$$

$$\frac{d(HX)}{dt} = 2k_1(H_2)(X_2) - 2k_{-1}(HX)^2$$

$$+\Big\{k_3k_4(H_2)(X_2) + k_4k_{-4}(X_2)(HX) - k_3k_{-3}(HX)(H_2) - k_{-3}k_{-4}(HX)^2$$

$$+ k_3k_{-3}(H_2)(HX) + k_3k_4(H_2)(X_2) - k_{-3}k_{-4}(HX)^2 - k_4k_{-4}(HX)(X_2)\Big\}$$

$$\times \frac{\sqrt{\dfrac{2k_2(X_2)}{k_{-2}}}}{k_{-3}(HX) + k_4(X_2)}$$

$$= 2k_1(H_2)(X_2) - 2k_{-1}(HX)^2 + \left[2k_3k_4(H_2)(X_2) - 2k_{-3}k_{-4}(HX)^2\right]$$

$$\times \frac{\sqrt{\dfrac{2k_2(X_2)}{k_{-2}}}}{k_{-3}(HX) + k_4(X_2)}$$

$$= 2k_1(H_2)(X_2)\left[1 - \frac{k_{-1}(HX)^2}{k_1(H_2)(X_2)}\right] + 2k_3k_4(H_2)(X_2)\left[1 - \frac{k_{-3}k_{-4}(HX)^2}{k_3k_4(H_2)(X_2)}\right]$$

$$\times \frac{\sqrt{\dfrac{2k_2(X_2)}{k_{-2}}}}{k_3(HX) + k_4(X_2)}$$

$$\frac{d(HX)}{dt} = 2k_1(H_2)(X_2)\left[1 - \frac{(HX)^2}{K(H_2)(X_2)}\right]\left[1 + \frac{k_3k_4}{k_1}\,\frac{\sqrt{\dfrac{2k_2(X_2)}{k_{-2}}}}{k_{-3}(HX) + k_4(X_2)}\right]$$

$$= 2k_1(H_2)(X_2)\left[1 - \frac{(HX)^2}{K(H_2)(X_2)}\right]\left[1 + \frac{\dfrac{k_3}{k_1}\sqrt{\dfrac{2k_2}{k_{-2}(X_2)}}}{1 + \dfrac{k_{-3}(HX)}{k_4(X_2)}}\right]$$

15.33 For the gas reaction

$$O + O_2 + M \underset{k'}{\overset{k}{\rightleftharpoons}} O_3 + M$$

When $M = O_2$ Benson and Axworthy [J. Chem Phys., 26, 1718 (1957)] obtained

$$k = 6.0 \times 10^7 e^{0.6/RT} \text{ L}^2 \text{ mol}^{-2} \text{ s}^{-1}$$

where the activation energy is in kcal mol^{-1}. Calculate the values of the parameters in the Arrhenius equation for the reverse reaction.

SOLUTION:

$$\Delta H° = 34.1 - 59.553 = -25.5 \text{ kcal mol}^{-1}$$

$$\Delta S° = 57.08 - 38.467 - 49.003 = -30.39 \text{ cal K}^{-1} \text{mol}^{-1}$$

$$K_p = e^{-\Delta H°/RT} e^{\Delta S°/R}$$

$$= e^{25,500/RT} e^{-30.39/R}$$

$$K_c = K_p \left(\frac{P°}{c° RT}\right)^{\Sigma \nu_i} = K_p \left(\frac{1}{24.46}\right)^{-1} \text{ L mol}^{-1} = \frac{k}{k'}$$

$$k' = \frac{k}{24.46 K_p}$$

$$k' = \frac{6 \times 10^7 \; e^{600/RT}}{24.46 \; e^{25,500/RT} \; e^{-30.39/R}}$$

$$= \left[\frac{6 \times 10^7}{24.46} e^{30.39R} \right] e^{(600-25,500)/RT}$$

$$= 1.1 \times 10^{13} e^{-24,900/RT} \quad \text{L mol}^{-1} \text{ s}^{-1}$$

15.34 (a) At 27 min $k = 0.0348 \text{ min}^{-1}$
At 60 min $k = 0.0347 \text{ min}^{-1}$

(b) 20.0 min

15.35 0.25 min^{-1} 2.77 min 4.00 min

15.36 (a) first (b) $3.59 \times 10^{-5} \text{ s}^{-1}$ (c) 0.354

15.37 $4.03 \times 10^{-9} \text{ min}^{-1}$

15.38 14.6 %

15.39 (a) $0.107 \text{ L mol}^{-1} \text{ s}^{-1}$ (b) 2850 s

15.40 (a) 221 s (b) 82.5 s

15.41 $40.9 \text{ atm}^{-1} \text{ s}^{-1}$

15.42 (a) 80 % (b) 67.2 % (c) 61.7 %

15.43 (a) 3.5 % (b) 9 %

15.45 $-\frac{d(H_2SeO_3)}{dt} = k(H_2SeO_3)(H^+)^2(I^-)^3$

15.46 $\frac{d(BrO_3^-)}{dt} = k'(SO_3^{2-})^2(SO_4^{2-})^3(H^+)(Br^-)$

15.47 0.00750 mol L^{-1}

15.48 (a) 395 s (b) 32,900 ft^3

15.49 $(n-1)kt = \frac{1}{(A)^{n-1}} - \frac{1}{(A)_0^{n-1}}$

$t_{1/2} = \frac{2^{n-1} - 1}{(n-1)k(A)_0^{n-1}}$

15.50

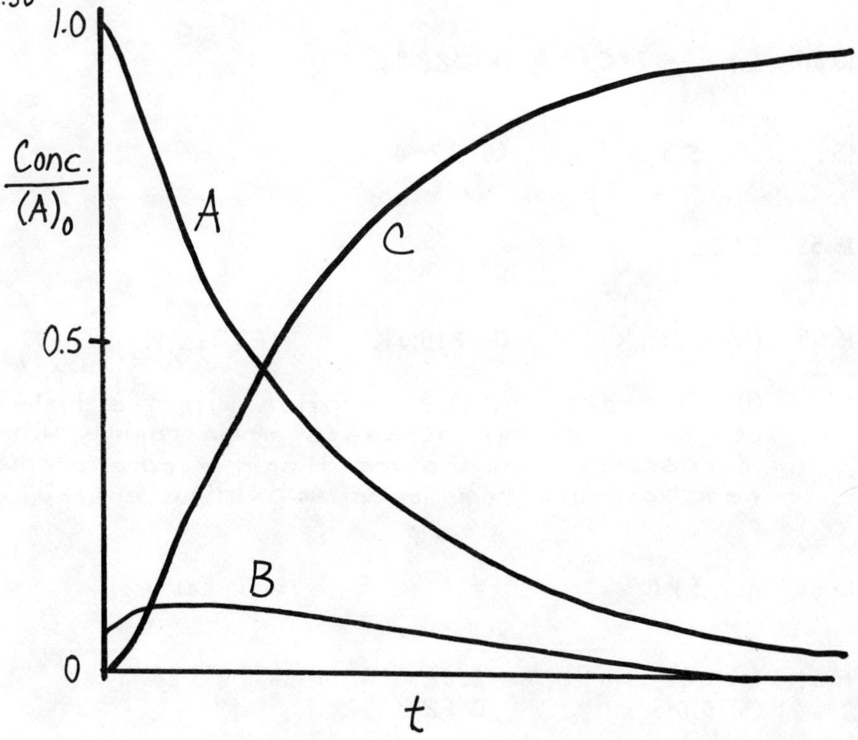

15.51 $(B) = k\,(A)_0\, t\, e^{-kt}$

$(C) = (A)_0\,[1 - e^{-kt}(1 + kT)]$

15.52 $\frac{d(O_2)}{dt} = 2k_1(NO)(O_3)$

15.53 $O_3 \rightleftharpoons O_2 + O$ (in equilibrium)

$O + O_3 \longrightarrow 2O_2$

15.54 $-\frac{d(A)}{dt} = \left(\frac{3}{4}k_1 + \frac{1}{4}\sqrt{k_1^2 + 8k_1 k_2 k_3/k_4}\right)(A)$

15.55 (a) 23.2 kcal mol^{-1} (b) $9.3 \times 10^{13}\ s^{-1}$ (c) 1.7 s

15.56 (a) 447°C (b) 388°C

15.57 (a) 5.3×10^{6} s (b) 2740 s

15.58 550 s

15.59 (a) 1091 K (b) 2180 K (c) 727 K

(d) The rate of the reaction with the higher activation energy increases more rapidly with increasing temperature than the rate of the reaction with the lower activation energy.

15.60 $A = 3 \times 10^{9}\ L^2\ mol^{-2}\ s^{-1}$ $E_a = 1370$ cal mol^{-1}

15.61 (a) 1.07 $atm^{-1}\ s^{-1}$ $2.06 \times 10^{-3}\ atm^{-1}\ s^{-1}$
(b) 0.013 s 0.62 s

15.62 (a) 18,700 cal mol^{-1} (b) 17,600 cal mol^{-1}
(c) 14.1 s^{-1} (d) 0.929 s^{-1}

15.63 (a) 2.0×10^{11} L mol^{-1} s^{-1} (b) 3.2×10^{-10} s
(c) 2.4×10^{-4} s

15.64 $10^{14.45}$ cm^3 mol^{-1} s^{-1}

15.66 -143, -130, and -125 cal K^{-1} mol^{-1} when the standard concentration is taken as 1 mol L^{-1}. The activated complex is more organized (lower entropy) than the reactants. This is especially true for the first reaction.

15.67 (a) 6.25×10^{14} s^{-1} (b) 9×10^{-7} s^{-1}

15.68 (a) 53.5 kcal mol^{-1} (b) 4.29 cal K^{-1} mol^{-1}

15.69 $K = (AB)/(A)(B) = \exp\left(\frac{\Delta S^\circ}{R}\right)\exp\left(-\frac{\Delta H^\circ}{RT}\right)$

$$\frac{d(D)}{dt} = k(C)(AB) = k(C)(A)(B)\exp\left(\frac{\Delta S^\circ}{R}\right)\exp\left(-\frac{\Delta H^\circ}{RT}\right)$$

$$= s\exp\left(-\frac{E_a}{RT}\right)(C)(A)(B)\exp\left(\frac{\Delta S^\circ}{R}\right)\exp\left(-\frac{\Delta H^\circ}{RT}\right)$$

$$k = A\exp\left[\frac{-E_a + H^\circ}{RT}\right]$$

where $A = s\exp\left(\frac{\Delta S^\circ}{R}\right)$

CHAPTER 16 Kinetics : Liquid Phase

16.1 Show that if A and B can be represented by spheres of the same radius that react when they touch the second-order rate constant is given by

$$k_a = \frac{8 \times 10^3 RT}{3\eta} \text{ L mol}^{-1} \text{ s}^{-1}$$

where R is in $J K^{-1} mol^{-1}$. To obtain this result D is expressed in terms of the radius of a spherical particle by use of equation 20.12. For water at 25°C, $\eta = 8.95 \times 10^{-4}$ kg m^{-1} s^{-1}. Calculate k at 25°C.

SOLUTION:

The diffusion coefficient for a spherical particle of radius r is given by

$$D = \frac{RT}{N_A 6\pi\eta r}$$

Subsituting in equation 16.4

$$k_a = 4\pi (10^3 \text{ L m}^{-3}) N_A (D_1 + D_2) R_{12}$$

$$= 4\pi (10^3 \text{ L m}^{-3}) \left(\frac{2RT}{6\pi\eta r}\right)(2r)$$

$$= \frac{8 \times 10^3 RT}{3\eta} \text{ L mol}^{-1} \text{ s}^{-1}$$

where RT is expressed in J mol^{-1} and η is expressed in kg m^{-1} s^{-1} to obtain the indicated units for the second-order rate constant.

For water at 25 °C

$$k_a = \frac{8(10^3\ L\ m^{-3})(8.314\ J\ K^{-1}\ mol^{-1})(298\ K)}{3(8.95 \times 10^{-4}\ kg\ m^{-1}\ s^{-1})}$$

$$= 7.4 \times 10^9\ L\ mol^{-1}\ s^{-1}$$

16.2 The diffusion coefficient D of an ion is related to its ionic mobility u by

$$D = \frac{uRT}{zF}$$

D = diffusion coefficient in $m^2\ s^{-1}$

u = ion mobility in $m^2\ V^{-1}\ s^{-1}$

R = 8.314 $J\ K^{-1}\ mol^{-1}$

z = number of charges on ion, and

F = Faraday constant = 96,500 C mol^{-1}

The ionic mobilities of H^+ and OH^- are $3.63 \times 10^{-7}\ m^2\ V^{-1}\ s^{-1}$ and $2.05 \times 10^{-7}\ m^2\ V^{-1}\ s^{-1}$ at 25°C. What is the rate constant for the reaction

$$H^+ + OH^- \longrightarrow H_2O$$

The reaction radius is 0.75 nm because once the proton is this close the reaction can proceed by quantum mechanical tunneling. The electrostatic factor f is 1.70.

SOLUTION:

$$D_1 = \frac{(3.63 \times 10^{-7}\ m^2\ V^{-1} s^{-1})(8.314 J K^{-1}\ mol^{-1})(298\ K)}{96,500 C\ mol^{-1}}$$

$$= 9.31 \times 10^{-9}\ m^2\ s^{-1}$$

$$D_2 = \frac{(2.05\times10^{-7}\ m^2\ V^{-1}s^{-1})(8.314\ J\ K^{-1}mol^{-1})(298\ K)}{96,500\ C\ mol^{-1}}$$

$$= 5.26\times10^{-9}\ m^2\ s^{-1}$$

$$k_a = 4\pi\ (10^3\ L\ m^{-3})\ N_A\ (D_1 + D_2)\ R_{12} f$$

$$= 4\pi\ (10^3\ L\ m^{-3})(6.022\times10^{23}\ mol^{-1})(14.57\times10^{-9}\ m^2 s^{-1})$$

$$\times(0.75\times10^{-9}\ m)\ 1.7$$

$$= 1.4\times10^{11}\ L\ mol^{-1}s^{-1}$$

16.3 For acetic acid at 25°C

$$CH_3CO_2H \underset{4.5\times10^{10}\ L\ mol^{-1}s^{-1}}{\overset{7.8\times10^5\ s^{-1}}{\rightleftharpoons}} CH_3CO_2^- + H^+$$

What is the relaxation time τ for a 0.1 mol L^{-1} solution?

SOLUTION:

$$K = \frac{(H^+)(CH_3CO_2^-)}{(CH_3CO_2H)} = \frac{7.8\times10^5\ s^{-1}}{4.5\times10^{10}\ L\ mol^{-1}s^{-1}} = 1.7\times10^{-5}\ mol\ L^{-1}$$

$$= \frac{(H^+)^2}{0.1-(H^+)}$$

$$(H^+) = 1.3\times10^{-3}\ mol\ L^{-1}$$

From equation 16.15

$$\tau = \{k_1 + k_1\ [(A)_{eq} + (B)_{eq}]\}^{-1}$$

$$= [7.8\times10^5 s^{-1} + (4.5\times10^{10}\ L\ mol^{-1}s^{-1})(2.6\times10^{-3}\ mol\ L^{-1})]^{-1}$$

$$= 8.5\ ns$$

16.5 Calculate the first-order rate constants for the dissociation of the following weak acids: acetic acid, acid form of imidazole $C_3N_2H_5^+$, NH_4^+. The corresponding acid dissociation constants are 1.75×10^{-5}, 1.2×10^{-7}, 5.71×10^{-10}, respectively. The second-order rate constants for the formation of the acid forms from a proton plus the base are 4.5×10^{10}, 1.5×10^{10}, and 4.3×10^{10} L mol^{-1} s^{-1} respectively.

SOLUTION:

For $HA \underset{k_2}{\overset{k_1}{\rightleftharpoons}} H^+ + A^-$

$$K = \frac{(H^+)(A^-)}{(HA)} = \frac{k_1}{k_2}$$

For acetic acid $k_1 = k_2 K$

$$= (4.5 \times 10^{10}\ \text{L mol}^{-1}\text{s}^{-1})(1.75 \times 10^{-5}\ \text{mol L}^{-1})$$

$$= 7.9 \times 10^{5}\ \text{s}^{-1}$$

For imidazole $k_1 = k_2 K$

$$= (1.5 \times 10^{10}\ \text{L mol}^{-1}\text{s}^{-1})(1.2 \times 10^{-7}\ \text{mol L}^{-1})$$

$$= 1.8 \times 10^{3}\ \text{s}^{-1}$$

For NH_4^+ $k_1 = k_2 K$

$$= (4.3 \times 10^{10}\ \text{L mol}^{-1}\text{s}^{-1})(5.71 \times 10^{-10}\ \text{mol L}^{-1})$$

$$= 25\ \text{s}^{-1}$$

16.6 The mutation of glucose is first order in glucose concentration and is catalyzed by acids (A) and bases (B). The first order rate constant may be expressed by an equation of the type that is encountered in reactions with parallel paths.

$$k = k_o + k(H^+) + k_A\,(A) + k_B\,(B)$$

where k_o is the first-order rate constant in the absence of acids and bases other than water. The following data were obtained by J. N. Bronsted and E. A. Guggenheim [J. Am. Chem. Soc., 49, 2554 (1927)] at 18 °C in a medium containing 0.02 M sodium acetate and various concentrations of acetic acid:

$(CH_3CO_2H)/\text{mol L}^{-1}$	0.020	0.105	0.199
$k/10^{-4}\ \text{min}^{-1}$	1.36	1.40	1.46

Calculate k_o and k_A. The term involving k_{H^+} is negligible under these conditions.

SOLUTION:

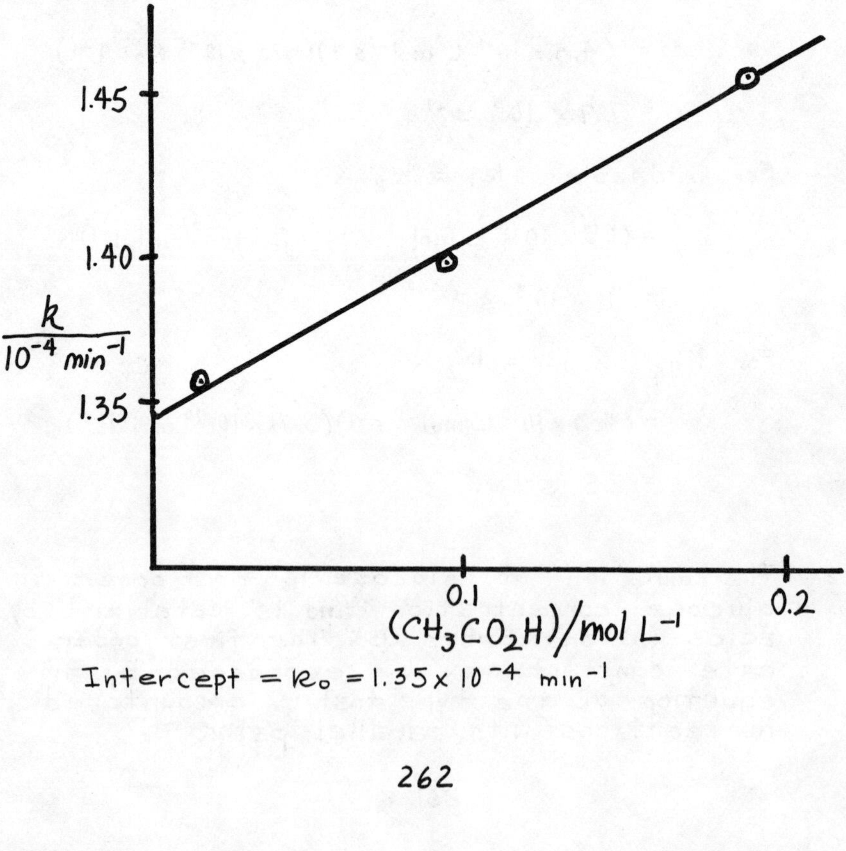

Intercept $= k_o = 1.35 \times 10^{-4}\ \text{min}^{-1}$

$$\text{slope} = \frac{(1.46 - 1.35) \times 10^{-4}}{0.2\ \text{mol L}^{-1}} = 5.5 \times 10^{-5}\ \text{L mol}^{-1}\text{min}^{-1} = k_A$$

16.7 The mutarotation of glucose is catalyzed by acids and bases and is first order in the concentration of glucose. When perchloric is used as a catalyst, the concentration of hydrogen ions may be taken to be equal to the concentration of perchloric acid, and the catalysis by perchlorate ion may be ignored since it is such a weak base. The following first-order constants were obtained by J. N. Bronsted and E. A. Guggenheim [J. Am. Chem. Soc., 49, 2554 (1927)] at 18°C:

$(HClO_4)/\text{mol L}^{-1}$	0.0010	0.0048	0.0099	0.0192	0.0300	0.0400
$k/10^{-4}\ \text{min}^{-1}$	1.25	1.38	1.53	1.90	2.15	2.59

Calculate the values of the constants in the equation $k = k_0 + k_{H^+}(H^+)$.

SOLUTION:

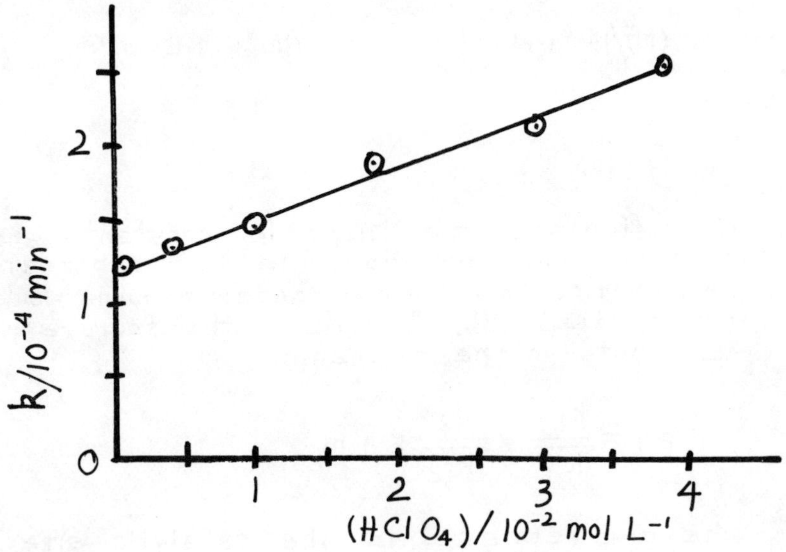

Intercept = k_0 = 1.21×10^{-4} min^{-1}

slope = k_{H^+} = 3.4×10^{-3} L mol^{-1} min^{-1}

16.9 The kinetics of the fumarase reaction

fumarate + H_2O = L-malate

is studied at 25°C using a 0.01 ionic strength buffer of pH 7. The rate of the reaction is obtained using a recording ultraviolet spectrometer to measure the fumarate concentration. The following rates of the forward reaction are obtained using a fumarase concentration of 5×10^{-10} mol L^{-1}.

(F)/10^{-6} mol L^{-1}	V_F/10^{-7} mol L^{-1} s^{-1}
2	2.2
40	5.9

The following rates of the reverse reaction are obtained using a fumarase concentration of 5×10^{-10} mol L^{-1}.

(M)/10^{-6} mol L^{-1}	V_M/10^{-7} mol L^{-1} s^{-1}
5	1.3
100	3.6

(a) Calculate the Michaelis constants and turnover numbers for the two substrates. In practice many more concentrations would be studied. (b) Calculate the four rate constants in the mechanism

$$E + F \underset{k_{-1}}{\overset{k_1}{\rightleftharpoons}} X \underset{k_{-2}}{\overset{k_2}{\rightleftharpoons}} E + M$$

where E represents the catalytic site.

There are four catalytic sites per fumarase molecule. (c) Calculate Keq for the reaction catalyzed. The concentration of H_2O is omitted in the expression for the equilibrium constant because its concentration cannot be varied in dilute aqueous solutions.

SOLUTION:

$$v_F = \frac{V_F}{1 + K_F/(F)}$$

$$v_F + \frac{v_F}{(F)} K_F = V_F$$

$$2.2 \times 10^{-7} + 0.110\ K_F = V_F$$

$$5.9 \times 10^{-7} + 0.0148\ K_F = V_F$$

$$-3.7 \times 10^{-7} + 0.0952\ K_F = 0$$

$$K_F = \frac{3.7 \times 10^{-7}}{0.0952} = 3.9 \times 10^{-6}\ \text{mol L}^{-1}$$

$$V_F = 2.2 \times 10^{-7} + (0.110)(3.9 \times 10^{-6}) = 6.5 \times 10^{-7}\ \text{mol L}^{-1}\ \text{s}^{-1}$$

$$v_M = \frac{V_M}{1 + K_M/(M)}$$

$$v_M + \frac{v_M}{(M)} K_M = V_M$$

$$1.3 \times 10^{-7} + 0.0260\ K_M = V_M$$

$$3.6 \times 10^{-7} + 0.0036\ K_M = V_M$$

$$-2.3 \times 10^{-7} + 0.0224\ K_M = 0$$

$$K_M = \frac{2.3 \times 10^{-7}}{0.0224} = 1.03 \times 10^{-5} \text{ mol L}^{-1}$$

$$V_M = 1.3 \times 10^{-7} + 0.026\,(1.03 \times 10^{-5}) = 4.0 \times 10^{-7} \text{ mol L}^{-1}\text{s}^{-1}$$

The rate constants should be expressed in terms of the concentration of enzymatic sites and so

$$k_2 = \frac{V_F}{(E)_0} = \frac{6.5 \times 10^{-7} \text{ mol L}^{-1} \text{ s}^{-1}}{4(5 \times 10^{-10} \text{ mol L}^{-1})} = 3.3 \times 10^{2} \text{ s}^{-1}$$

$$k_{-1} = \frac{V_M}{(E)_0} = \frac{4.0 \times 10^{-7} \text{ mol L}^{-1} \text{ s}^{-1}}{4(5 \times 10^{-10} \text{ mol L}^{-1})} = 2.0 \times 10^{2} \text{ s}^{-1}$$

(b) $$K_F = \frac{k_2 + k_1}{k_1}$$

$$k_1 = \frac{k_2 + k_{-1}}{K_F} = \frac{(3.3 + 2.0) \times 10^{2} \text{ s}^{-1}}{3.9 \times 10^{-6} \text{ mol L}^{-1}} = 1.4 \times 10^{8} \text{ L mol}^{-1} \text{s}^{-1}$$

$$K_M = \frac{k_2 + k_1}{k_{-2}}$$

$$k_{-2} = \frac{k_2 + k_{-1}}{K_M} = \frac{(3.3 + 2.0) \times 10^{2} \text{ s}^{-1}}{1.03 \times 10^{-5} \text{ mol L}^{-1}}$$

$$= 5.1 \times 10^{7} \text{ L mol}^{-1} \text{ s}^{-1}$$

(c) $$K = \frac{(M)_{eq}}{(F)_{eq}} = \frac{V_F K_M}{V_M K_F}$$

$$= \frac{(6.5 \times 10^{-7})\,(1.03 \times 10^{-5})}{(4.0 \times 10^{-7})\,(3.9 \times 10^{-6})} = 4.3$$

$$K = \frac{k_1 k_2}{k_{-1} k_{-2}}$$

$$= \frac{(1.4 \times 10^8)(3.3 \times 10^2)}{(2.0 \times 10^2)(5.1 \times 10^7)} = 4.5$$

16.11 At 25°C and pH 8 the maximum initial velocities V and Michaelis constants K of the fumarase reaction F + H_2O = M are

$V_F = (0.2 \times 10^3\ s^{-1})(E)_o$ $\qquad$ $V_M = (0.6 \times 10^3\ s^{-1})(E)_o$

$K_F = 7 \times 10^{-6}$ mol L^{-1} $\qquad$ $K_M = 100 \times 10^{-6}$ mol L^{-1}

where $(E)_o$ is the total molar concentration of enzymatic sites. Calculate the values of the four rate constants in the mechanism

$$E + F \underset{k_2}{\overset{k_1}{\rightleftharpoons}} EX \underset{k_4}{\overset{k_3}{\rightleftharpoons}} E + M$$

and the equilibrium constant $(M)_{eq}/(F)_{eq}$.

SOLUTION:

$V_F = k_3\ (E)_o$ $\qquad$ $K_F = \dfrac{k_2 + k_3}{k_1}$

$V_M = k_2\ (E)_o$ $\qquad$ $K_M = \dfrac{k_2 + k_3}{k_4}$

$k_3 = 0.2 \times 10^3\ s^{-1}$

$k_2 = 0.6 \times 10^3\ s^{-1}$

$$k_1 = \frac{k_2 + k_3}{K_F} = \frac{0.8 \times 10^3\ s^{-1}}{7 \times 10^{-6}\ mol\ L^{-1}} = 1.1 \times 10^8\ L\ mol^{-1} s^{-1}$$

$$k_4 = \frac{0.8 \times 10^3\ s^{-1}}{100 \times 10^{-6}\ mol\ L^{-1}} = 8 \times 10^6\ L\ mol^{-1}\ s^{-1}$$

$$K_{eq} = \frac{(M)_{eq}}{(F)_{eq}} = \frac{V_F K_M}{V_M K_F} = \frac{k_1 k_3}{k_2 k_4}$$

$$= 4.6$$

16.12 Derive the steady-state rate equation for the mechanism

$$E + S \underset{k_2}{\overset{k_1}{\rightleftharpoons}} X \xrightarrow{k_3} E + P$$

$$E + I \underset{k_5}{\overset{k_4}{\rightleftharpoons}} EI$$

for the case that $(S) \gg (E)_0$ and $(I) \gg (E)_0$.

SOLUTION:

$$(E)_0 = (E) + (EI) + (X)$$

$$= (E)[1 + (I)/K_I] + (X) \qquad (1)$$

since

$$K_I = \frac{(E)(I)}{(EI)} = \frac{k_5}{k_4}$$

Assuming that X is in a steady state

$$\frac{d(X)}{dt} = k_1(E)(S) - (k_2 + k_3)(X) = 0 \qquad (2)$$

Solving equation 1 for (E) and substituting this expression in equation 2 yields

$$\frac{k_1(S)(E)_0}{1 + (I)/K_I} = (X)\left[\frac{k_1(S)}{1 + (I)/K_I} + k_2 + k_3\right]$$

$$\frac{d(P)}{dt} = k_3(X) = \frac{k_3(E)_0}{1 + \frac{k_2 + k_3}{k_1(S)}\left[1 + \frac{(I)}{K_I}\right]}$$

$$= \frac{V_S}{1 + \frac{K_S}{(S)}\left[1 + \frac{(I)}{K_I}\right]}$$

16.13 The following initial velocities were determined spectrophotometrically for solutions of sodium succinate to which a constant amount of succinoxidase was added. The velocities are given as the change in absorbancy at 250 nm in 10 s. Calculate V, Ks, and K_I for malonate.

(succ.) 10^{-3} mol L^{-1}	$\frac{A \times 10^3}{10 s}$ No. Inhib.	15×10^{-6} mol L^{-1} malonate
10	16.7	14.9
2	14.2	10.0
1	11.3	7.7
0.5	8.8	4.9
0.33	7.1	—

SOLUTION:

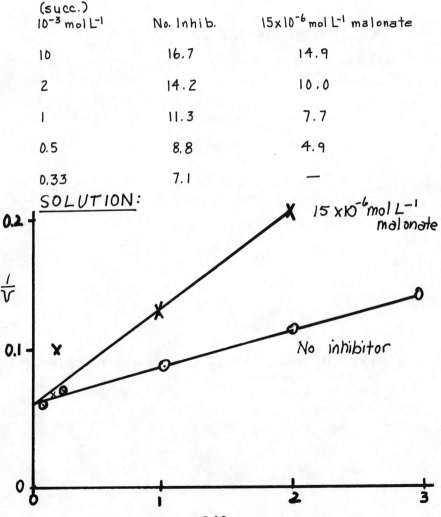

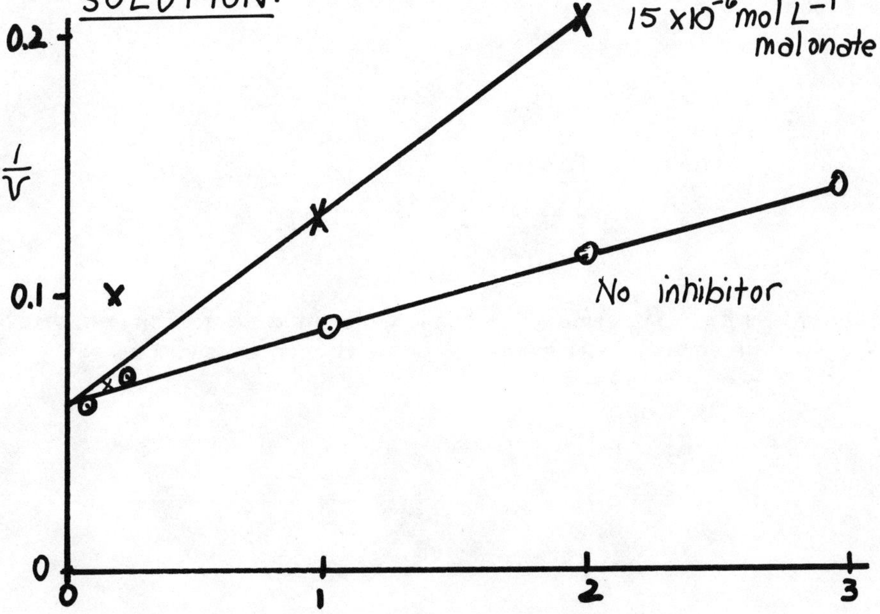

Equation 16.42

$$v = \frac{d(P)}{dt} = \frac{V_S}{1 + \frac{K_S}{(S)}\left[1 + \frac{(I)}{K_I}\right]}$$

$$\frac{1}{v} = \frac{1}{V_S} + \frac{K_S}{V_S(S)}\left[1 + \frac{(I)}{K_I}\right]$$

$$\frac{1}{V_S} = 0.0662 \qquad V_S = 15.1$$

Slope without inhibitor

$$= \frac{0.141 - 0.066}{3 \times 10^3 \text{ L mol}^{-1}} = 25 \times 10^{-6} = \frac{K_S}{V_S}$$

$$K_S = (25 \times 10^{-6})(15.1) = 0.38 \times 10^{-3} \text{ mol L}^{-1}$$

Slope with inhibitor

$$= \frac{0.204 - 0.066}{2 \times 10^3 \text{ L mol}^{-1}} = 0.069 = \frac{K_S}{V_S}\left[1 + \frac{(I)}{K_I}\right]$$

$$0.069 = \frac{0.38}{15.1}\left[1 + \frac{15 \times 10^{-6} \text{ mol L}^{-1}}{K_I}\right]$$

$$K_I = \frac{15 \times 10^{-6} \text{ mol L}^{-1}}{\frac{(15.1)(0.069)}{0.38} - 1} = 8.6 \times 10^{-6} \text{ mol L}^{-1}$$

16.14 The maximum initial velocities for an enzymatic reaction are determined at a series of pH values

pH	V
6.0	11
6.5	30

7.0	74
7.5	129
8.0	147
8.5	108
9.0	53

Calculate the values of the parameters V', K_a and K_b in

$$V = \frac{V'}{1 + (H^+)/K_a + K_b/(H^+)}$$

Hint : A plot of V versus pH may be constructed and the hydrogen ion concentration at the midpoint on the acid side referred to as $(H^+)_a$ and the hydrogen ion concentration at the midpoint on the basic side is referred to as $(H^+)_b$. Then

$$K_a = (H^+)_a + (H^+)_b - 4\sqrt{(H^+)_a (H^+)_b}$$

$$K_b = \frac{(H^+)_a (H^+)_b}{K_a}$$

SOLUTION:

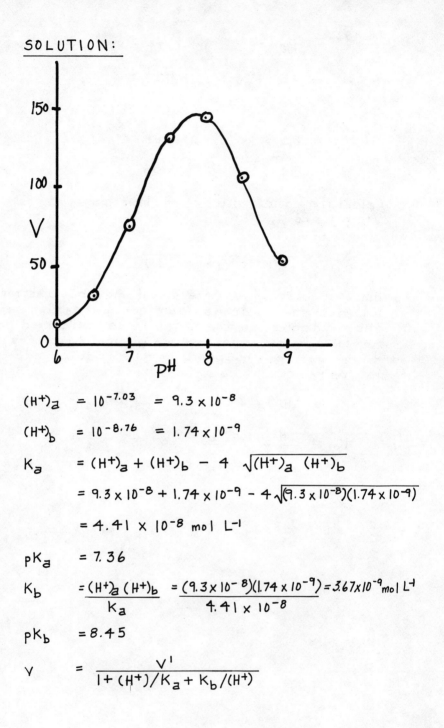

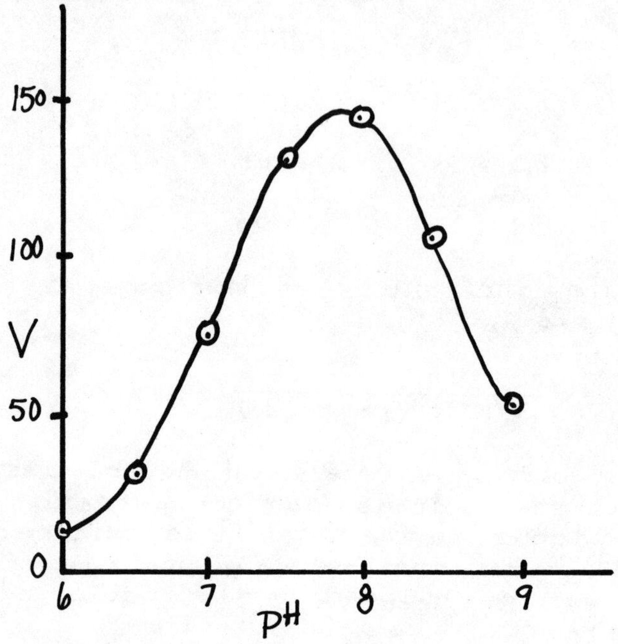

$(H^+)_a = 10^{-7.03} = 9.3 \times 10^{-8}$

$(H^+)_b = 10^{-8.76} = 1.74 \times 10^{-9}$

$K_a = (H^+)_a + (H^+)_b - 4\sqrt{(H^+)_a\,(H^+)_b}$

$= 9.3 \times 10^{-8} + 1.74 \times 10^{-9} - 4\sqrt{(9.3 \times 10^{-8})(1.74 \times 10^{-9})}$

$= 4.41 \times 10^{-8}\ \text{mol L}^{-1}$

$pK_a = 7.36$

$K_b = \dfrac{(H^+)_a\,(H^+)_b}{K_a} = \dfrac{(9.3 \times 10^{-8})(1.74 \times 10^{-9})}{4.41 \times 10^{-8}} = 3.67 \times 10^{-9}\ \text{mol L}^{-1}$

$pK_b = 8.45$

$v = \dfrac{V'}{1 + (H^+)/K_a + K_b/(H^+)}$

$$147 = \frac{V'}{1 + (10^{-8})/(4.41 \times 10^{-8}) + (0.367 \times 10^{-8})/(10^{-8})}$$

$$V' = 234$$

16.16 What current density i will be obtained for the evolution of hydrogen gas on (a) Pt and (b) Fe from 1 mol L^{-1} HCl at 25°C at an overpotential of 0.1 volt?

SOLUTION:

$$\text{(a)} \quad i = i_0 e^{E_{\hat{\eta}}/2.303a}$$

$$= (10^{-2.6} \text{ A cm}^{-2}) e^{0.1/(2.303)(0.028)}$$

$$= 0.012 \text{ A cm}^{-2}$$

$$\text{(b)} \quad i = (10^{-6} \text{ A cm}^{-2}) e^{0.1/(2.303)(0.130)}$$

$$= 1.4 \times 10^{-6} \text{ A cm}^{-2}$$

16.17 0.6 nm

16.18 2.7×10^{-4} s

16.19 $k_1 = 1.8 \times 10^{10}$ L mol^{-1} s^{-1} $\quad k_{-1} = 1.1 \times 10^3$ s^{-1}

16.20 $1/\tau = k_1 [(A)_{eq} + (B)_{eq}] + k_2 [(C)_{eq} + (D)_{eq}]$

16.21 $1/\tau = k_1 + k_1' K_A + k_{-1} + k_{-1}' K_B$

16.22 H^+ production : $k = 1.75 \times 10^5, 1.2 \times 10^2, 5.71\ s^{-1}$

OH^- production : $k = 5.71, 0.83 \times 10^3, 1.8 \times 10^5\ s^{-1}$

Imidazole can play both roles about equally well.

16.23 $K_1 = (HOCl)(OH^-)/(OCl^-)$

$$\frac{d(Cl^-)}{dt} = k(I^-)(HOCl) = k(I^-)K_1(OCl^-)/(OH^-)$$

16.24 $V = 1.2 \times 10^{-6}\ mol\ L^{-1}\ s^{-1}$ $\quad K_S = 0.48 \times 10^{-3}\ mol\ L^{-1}$

16.25 (a) $k_1 = 1.3 \times 10^8\ L\ mol^{-1} s^{-1}$ $\quad k_{-1} = 0.2 \times 10^3\ s^{-1}$

$k_2 = 0.33 \times 10^3\ s^{-1}$ $\quad k_{-2} = 5.3 \times 10^7\ L\ mol^{-1} s^{-1}$

(b) -820 cal mol^{-1}

16.26 $k_1 = 4.9 \times 10^8\ L\ mol^{-1} s^{-1}$ $\quad k_{-1} = 2.6 \times 10^3\ s^{-1}$

$k_2 = 0.8 \times 10^3\ s^{-1}$ $\quad k_{-2} = 3.4 \times 10^7\ L\ mol^{-1}\ s^{-1}$

16.27 $$\frac{d(P_2)}{dt} = \frac{[k_1 k_2 (S) - k_{-1} k_{-2} (P_1)(P_2)]\ (E)_0}{k_1 (S) + k_{-1}(P_1) + k_2 + k_{-2}(P_2)}$$

$$v_f = \frac{k_2 (E)_0}{1 + \dfrac{k_2}{k_1 (S)}}$$

$$v_r = \frac{k_{-1}(P_1)(E)_0}{1 + \dfrac{k_2}{k_{-2}(P_2)} + \dfrac{k_{-1}(P_1)}{k_{-2}(P_2)}}$$

Thus all four rate constants can be determined.

16.28 $V = k_2 k_3 (E)_0 / (k_1 + k_3)$ $K_S = k_2 / (k_1 + k_3)$

Since no enzyme-substrate complex is formed, this mechanism does not account for the selectivity of enzyme-catalyzed reactions.

16.29 The product completely inhibits the enzyme, putting it out of action before all the S is used up. Since the binding of product is reversible, the reaction can be restarted by adding more substrate and displacing the product.

16.30 33 %

16.31 $V_S = k_2 (E)_0 / [1 + (H^+)/K_{EHS}]$

$$K_S = \frac{k_{-1} + k_2}{k_1} \quad \frac{[1 + (H^+)/K_{EH}]}{[1 + (H^+)/K_{EHS}]}$$

16.32 $$V = \frac{k_2 (alc) (E)_0}{1 + \dfrac{k_2 (alc)}{k_1 (NAD^+)}}$$

16.33 (a) 0.045, (b) 1.31 V

CHAPTER 17 Photochemistry

17.1 A certain photochemical reaction requires an activation energy of 30 Kcal mol^{-1}. To what values does this correspond in the following units : (a) J per mole, (b) frequency of light, (c) wave number, (d) wavelength in nm, (e) electron volts?

SOLUTION:

(a) $(30 \text{ Kcal mol}^{-1})(4.184 \text{ KJ Kcal}^{-1}) = 126 \text{ KJ mol}^{-1}$

(b) $$\nu = \frac{E}{h} = \frac{1.26 \times 10^{5} \text{ J mol}^{-1}}{(6.63 \times 10^{-34} \text{ Js})(6.02 \times 10^{23} \text{ mol}^{-1})}$$

$$= 3.16 \times 10^{14} \text{ s}^{-1}$$

(c) $$\tilde{\nu} = \frac{\nu}{c} = \frac{(3.16 \times 10^{14} \text{ s}^{-1})(10^{-2} \text{ m cm}^{-1})}{2.998 \times 10^{8} \text{ ms}^{-1}}$$

$$= 10{,}500 \text{ cm}^{-1}$$

(d) $$\lambda = \frac{1}{\tilde{\nu}} = \frac{1}{10{,}500 \text{ cm}^{-1}} = 9.52 \times 10^{-5} \text{ cm}$$

$$= 925 \text{ nm}$$

(e) $$\frac{30 \text{ Kcal mol}^{-1}}{23.060 \text{ Kcal mol}^{-1} \text{ eV}^{-1}} = 1.30 \text{ eV}$$

17.2 A sample of gaseous acetone is irradiated with monochromatic light having a wavelength of 313 nm. Light of this wavelength decomposes the acetone according to the equation.

$$(CH_3)_2CO \longrightarrow C_2H_6 + CO$$

The reaction cell used has a volume of 59 cm^3. The acetone vapor absorbs 91.5% of the incident energy. During the experiment the following data are obtained:

Temperature of reaction	= 56.7°C
Initial pressure	= 766.3 torr
Final pressure	= 783.2 torr
Time of radiation	= 7 hr.
Incident energy	$= 48.1 \times 10^{-4}\ J\,s^{-1}$

What is the quantum yield?

SOLUTION:

Since $PV = nRT$, $V\Delta P = RT\Delta n$

$$\Delta n = \frac{V\Delta P}{RT} = \frac{(0.059\ L)(783.2\ torr - 766.3\ torr)}{(760\ torr\ atm^{-1})(0.082\ L\ atm\ K^{-1} mol^{-1})(329.85K)}$$

$$= 4.85 \times 10^{-5}\ mol$$

$$= (4.85 \times 10^{-5})(6.02 \times 10^{23}\ mol^{-1})$$

$$= 2.92 \times 10^{19}\ \text{molecules reacting}$$

$$E = \frac{hc}{\lambda} = \frac{(6.626 \times 10^{-34} Js)(2.998 \times 10^{8}\ m\,s^{-1})}{313 \times 10^{-9}\ m}$$

$$= 6.36 \times 10^{-19}\ J$$

Number of quanta absorbed

$$= \frac{(48.1 \times 10^{-4}\ J\,s^{-1})(7 \times 60 \times 60\,s)(0.915)}{6.36 \times 10^{-19}\ J}$$

$$= 1.75 \times 10^{20}$$

$$\Phi = \frac{2.92 \times 10^{19}}{1.75 \times 10^{20}} = 0.167$$

17.3 A 100 cm^3 vessel containing hydrogen and chlorine was irradiated with light of 400 nm. Measurements with a thermopile showed that 11×10^{-7} J of light energy was absorbed by the chlorine per second. During an irradiation of 1 min the partial pressure of chlorine, as determined by the absorption of light and the application of Beer's law, decreased from 205 to 156 torr (corrected to 0°C). What is the quantum yield?

SOLUTION:

$$E = \frac{hc}{\lambda} = \frac{(6.626 \times 10^{-34}\ \text{J s})(2.998 \times 10^{8}\ \text{m s}^{-1})}{400 \times 10^{-9}\ \text{m}}$$

$$= 4.97 \times 10^{-19}\ \text{J}$$

Number of quanta absorbed

$$= \frac{(11 \times 10^{-7}\ \text{J s}^{-1})(60\ \text{s})}{4.97 \times 10^{-19}\ \text{J}}$$

$$= 1.33 \times 10^{14}$$

Since $PV = nRT$, $V\Delta P = RT\Delta n$

$$\Delta n = \frac{V\Delta P}{RT}$$

$$= \frac{(0.1\ \text{L})(205\ \text{torr} - 156\ \text{torr})(6.022 \times 10^{23}\ \text{mol}^{-1})}{(760\ \text{torr atm}^{-1})(0.082\ \text{L atm K}^{-1}\ \text{mol}^{-1})(273\ \text{K})}$$

$$= 1.73 \times 10^{20}$$

Since $H_2 + Cl_2 = 2\ HCl$

$$\Phi = \frac{1.73 \times 10^{20}}{1.33 \times 10^{14}} \times 2 = 2.6 \times 10^{6}\ \text{mol HCl einstein}^{-1}$$

17.5 For 900 s, light of 436 nm was passed into a carbon tetrachloride solution containing bromine and cinnamic acid. The average absorbed was $19.2 \times 10^{-4}\ J\ s^{-1}$. Some of the bromine reacted to give cinnamic acid dibromide, and in this experiment the total bromine content decreased by 3.83×10^{19} molecules. (a) What was the quantum yield? (b) State whether or not a chain reaction was involved. (c) If a chain mechanism was involved, suggest suitable reactions which might explain the observed quantum yield.

SOLUTION:

(a) $$E = \frac{hc}{\lambda} = \frac{(6.63 \times 10^{-34}\ Js)(3 \times 10^{8}\ ms^{-1})}{(436 \times 10^{-9}\ m)}$$

$$= 4.56 \times 10^{-19}\ J$$

$$\frac{19.2 \times 10^{4}\ J\ s^{-1}}{4.56 \times 10^{-19}\ J} = 4.21 \times 10^{15}\ s^{-1};$$ rate quanta are absorbed

$$(4.21 \times 10^{15}\ s^{-1})(900\ s) = 3.79 \times 10^{18}\ \text{quanta}$$

$$\Phi = \frac{3.83 \times 10^{19}}{3.79 \times 10^{18}} = 10.1$$

(b) Since $\Phi > 1$, a chain reaction is involved

(c) $$\phi CH = CHCO_2H + h\nu = \phi CH = CHCO_2H^*$$

$$\phi CH = CHCO_2H^* + Br_2 = \phi CHBrCHCO_2H^* + Br^*$$

$$\phi CHBrCHCO_2H^* + Br_2 = \phi CHBrCHBrCO_2H + Br^*$$

$$\phi CH = CHCO_2H + Br^* = CHBrCHCO_2H^*$$

17.6 Go through the steps of deriving equation 17.30 from the mechanism of Fig. 17.1 plus reaction 17.27.

SOLUTION:

With the addition of reaction 17.27, equation 17.8 becomes

$$(T_1) = \frac{k_{isc} I}{[k'_{isc} + k_p + k_r (R)][k_{ic} + k_f + k_{isc}]}$$

Substituting this equation 17.29 yields

$$\frac{1}{\Phi} = \frac{I}{k_r (T_1)(R)} = \frac{[k_{ic} + k_f + k_{isc}][k_p + k_{isc} + k_r (R)]}{k_r (R) k_{isc}}$$

$$= \frac{k_f + k_{ic} + k_{isc}}{k_{isc}} \left[1 + \frac{k_p + k_{isc}}{k_r (R)}\right]$$

17.7 A solution of a dye is irradiated with 400 nm light to produce a steady concentration of triplet state molecules. If the triplet state yield is 0.9, and the triplet state lifetime is 20×10^{-6} s, what light intensity, expressed in watts, is required to maintain a steady triplet concentration of 5×10^{-6} mol L^{-1} in a liter of solution. Assume that all of the light is absorbed.

SOLUTION:

$$\frac{d(T_1)}{dt} = 0.9 I - \frac{1}{20 \times 10^{-6}} (T_1) = 0$$

$$I = \frac{(5 \times 10^{-6} \text{ mol L}^{-1})(1 \text{ L})}{(20 \times 10^{-6} \text{ s})(0.9)} = 0.278 \text{ einstein s}^{-1}$$

$$E = \frac{N_A hc}{\lambda} = \frac{(6.022 \times 10^{23} \text{ mol}^{-1})(6.62 \times 10^{-34} \text{ J s})(3 \times 10^8 \text{ m s}^{-1})}{400 \times 10^{-9} \text{ m}}$$

$$= 2.99 \times 10^5 \text{ J einstein}^{-1}$$

$$I = (0.278 \text{ einstein s}^{-1})(2.99 \times 10^5 \text{ J einstein}^{-1})$$

$$= 83 \text{ kW}$$

17.8 The quantum yield is 2 for the photolysis of gaseous HI to $H_2 + I_2$ by light of 253.7 nm wavelength. Calculate the number of moles of HI that will be decomposed if 300 J of light of this wavelength is absorbed.

SOLUTION:

$$E = \frac{hc}{\lambda} = \frac{(6.626 \times 10^{-34} \text{ J s})(3 \times 10^8 \text{ m s}^{-1})}{(253.7 \times 10^{-9} \text{ m})}$$

$$= 7.84 \times 10^{-19} \text{ J}.$$

$$n = \frac{(2)(300 \text{ J})}{(7.84 \times 10^{-19} \text{ J})(6.02 \times 10^{23} \text{ mol}^{-1})} = 1.27 \times 10^{-3} \text{ mol of HI}$$

17.9 The following calculations are made on a uranyl oxalate actinometer, on the assumption that the energy of all wavelengths between 254 and 435 nm is completely absorbed. The actinometer contains 20 cm^3 of 0.05 mol L^{-1} oxalic acid, which also is 0.01 mol L^{-1} with respect to uranyl sulfate. After 2 hr of exposure to ultraviolet light, the solution required 34 cm^3 of potassium permanganate, $KMnO_4$, solution to titrate the undecomposed oxalic acid. The same volume, 20 cm^3, of unilluminated solution required 40 cm^3 of the $KMnO_4$ solution. If the average energy of the quanta in this range may be taken as corresponding to a wavelength of 350 nm, how many joules were absorbed per second in this experiment? (Given $\Phi = 0.57$)

SOLUTION:

Moles of oxalic acid decomposed

$$= (0.02\,L)(0.05\text{ mol }L^{-1})\left(\frac{40-34}{40}\right)$$

$$= 1.5 \times 10^{-4}\text{ mol}$$

Number of einsteins s^{-1} required

$$= \frac{(1.5 \times 10^{-4}\text{ mol})}{(2 \times 60 \times 60\,s)(0.57)}$$

$$= 3.65 \times 10^{-8}\text{ einsteins } s^{-1}$$

$$E = \frac{hc}{\lambda} = \frac{(6.626 \times 10^{-34}\,J\,s)(2.998 \times 10^{8}\,m\,s^{-1})}{3.5 \times 10^{-7}\,m}$$

$$= 5.68 \times 10^{-19}\,J$$

Energy flux

$$= (3.65 \times 10^{-8}\text{ einsteins } s^{-1})(6.022 \times 10^{23}\,mol^{-1}) \times (5.68 \times 10^{-19}\,J)$$

$$= 125 \times 10^{-4}\,J\,s^{-1}$$

(Note that einsteins and mole cancel because the einstein is a special name for a mole of photons.)

17.11 Keytone A dissolved in t-butyl alcohol is excited to a triplet state A* by light of 320-380 nm. The triplet may then return to the ground state A or rearrange to give the isomer B. Thus

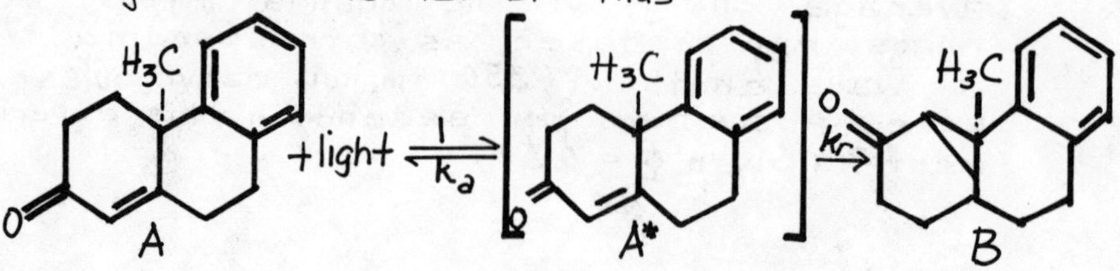

The unimolecular rate constant K_d determines the rate of deactivation of excited molecules and the unimolecular rate constant k_r determines the rate of rearrangement. The symbol I represents the number of einsteins per second and it is assumed that each photon absorbed produces one excited molecule of A*. The quantum yield Φ for the formation of B is given by

$$\Phi = \frac{d(B)/dt}{I} = \frac{k_r\,(A^*)}{I}$$

Zimmerman, McCullough, Staley and Padwa measured the quenching effect of dissolved napthalene and report the following data:

Moles napthalene liter^{-1}	0.0099	0.0330	0.0620	0.0680	0.0775	0.0960
Φ	0.0049	0.0023	0.0020	0.0017	0.0014	0.0012

The quenching rate by napthalene is controlled by the bimolecular quenching constant of A which is equal to the diffusion-controlled constant $k_q = 1.2 \times 10^9$ L mol^{-1} s^{-1}. Assuming a steady state.

$$d(A^*)/dt = I - k_r\,(A^*) - k_d\,(A^*) - k_q\,(A^*)(N) = 0$$

where N is the concentration of napthalene. Calculate k_r and k_d by plotting $1/\Phi$ versus (N) and determining the slope and intercept of the line.

SOLUTION

$$\frac{I}{(A^*)} = k_r + k_d + k_q\,(N)$$

$$\frac{1}{\Phi} = \frac{I}{k_r(A^*)} = 1 + \frac{k_d}{k_r} + \frac{k_q}{k_r}\,(N)$$

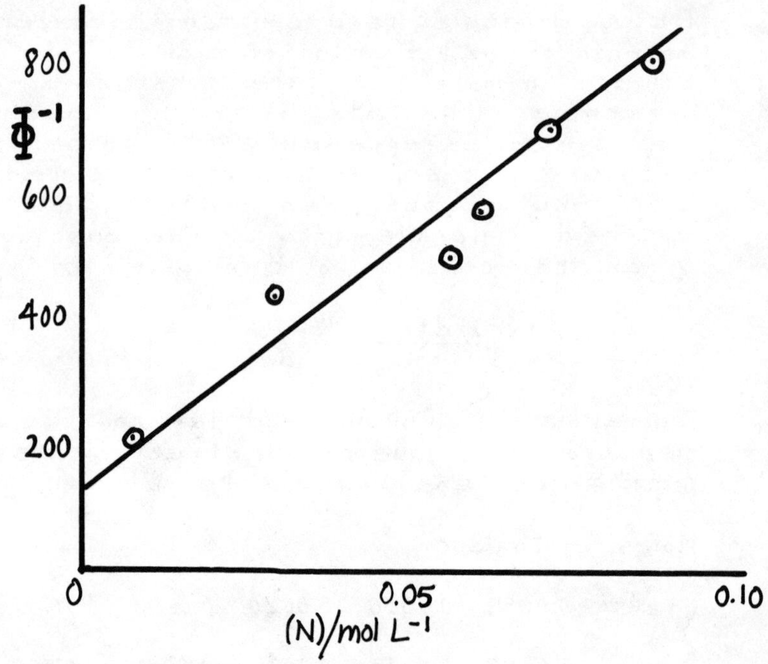

$$\text{Slope} = \frac{875 - 120}{0.1} = 7550 \text{ L mol}^{-1} = \frac{1.2 \times 10^9 \text{ L mol}^{-1}\text{ s}^{-1}}{k_r}$$

$$k_r = 1.6 \times 10^5 \text{ s}^{-1}$$

$$\text{Intercept} = 120 = 1 + \frac{k_d}{k_r} \quad 1 + \frac{k_d}{1.6 \times 10^5 \text{ s}^{-1}}$$

$$k_d = 1.9 \times 10^7 \text{ s}^{-1}$$

17.12 The photochemical oxidation of phosgene, sensitized by chlorine, has been studied by G.K. Rollefson and C.W. Montgomery [J. Am. Chem. Soc., 55, 142, 4025 (1932)]. The overall reaction is

$$2\,COCl_2 + O_2 = 2\,CO_2 + 2\,Cl_2$$

and the rate expression which gives the

effect of the several variables is

$$\frac{dc_{CO_2}}{dt} = \frac{k I_o (COCl_2)}{1 + k' (Cl_2)/(O_2)}$$

where I_o is the intensity of the light. The quantum yield is about two molecules per quantum. Devise a series of chemical equations involving the existence of the free radicals ClO and COCl which will give a mechanism with the rate expression.

SOLUTION:

$$COCl_2 + h\nu \xrightarrow{k_1} COCl + Cl$$

$$COCl + O_2 \xrightarrow{k_2} CO_2 + ClO$$

$$COCl_2 + ClO \xrightarrow{k_3} CO_2 + Cl_2 + Cl$$

$$COCl + Cl_2 \xrightarrow{k_4} COCl_2 + Cl$$

$$Cl + Cl \longrightarrow Cl_2$$

$$2\ COCl_2 + O_2 = 2\ CO_2 + 2\ Cl_2$$

In the steady state

$$\frac{d(COCl)}{dt} = I_o (COCl_2) - k_2 (COCl)(O_2) - k_4 (COCl)(Cl_2) = 0$$

Therefore $(COCl) = \dfrac{I_o (COCl_2)}{k_2 (O_2) + k_4 (Cl_2)}$

$$\frac{d(ClO)}{dt} = k_2 (COCl)(O_2) - k_3 (COCl)(ClO) = 0$$

Therefore $k_2 (COCl)(O_2) = k_3 (COCl)(ClO)$

$$\frac{d(CO_2)}{dt} = k_2 (COCl)(O_2) + k_3 (COCl_2)(ClO)$$

$$= 2k_2(COCl)(O_2)$$

$$= \frac{2I_0(COCl_2)}{1 + \frac{k_4(Cl_2)}{k_2(O_2)}}$$

17.14 Calculate the longest wavelength that can theoretically decompose water at 25°C in a one photon process to give H_2 (g) and $1/2\, O_2$ (g) in their standard states. Given: $\Delta G° = 56,690$ cal mol^{-1} for $H_2O(\ell) = H_2(g) + 1/2\, O_2(g)$.

SOLUTION:

$$\Delta G° = N_A hc/\lambda$$

$$\lambda = \frac{N_A hc}{\Delta G°} = \frac{(6.022 \times 10^{23}\,\text{mol}^{-1})(6.626 \times 10^{-34}\,\text{Js})(2.998 \times 10^{8}\,\text{m s}^{-1})}{(56,690\,\text{cal mol}^{-1})(4.184\,\text{J cal}^{-1})}$$

$$= 504\ \text{nm}$$

17.15 If a good agricultural crop yields about 2 tons acre^{-1} of dry organic material per year with a heat of combustion of about 4000 cal g^{-1}, what fraction of a year's solar energy is stored in an agricultural crop if the solar energy is about 1000 cal min^{-1} ft^{-2} and the sun shines about 500 min day^{-1} on the average? 1 acre = 43,560 ft^2, and 1 ton = 907,000 g.

SOLUTION:

$$\text{solar energy} = (43,560\,\text{ft}^2\,\text{acre}^{-1})(500\,\text{min day}^{-1}) \times (365\,\text{day year}^{-1})(10^3\,\text{cal ft}^{2}\,\text{min}^{-1})$$

$$= 7.96 \times 10^{12}\ \text{cal acre}^{-1}\ \text{year}^{-1}$$

$$\text{energy stored} = (2 \text{ ton acre}^{-1} \text{year}^{-1})(907{,}000 \text{ g ton}^{-1}) \times (4 \times 10^3 \text{ cal g}^{-1})$$

$$= 7.26 \times 10^9 \text{ cal acre}^{-1} \text{ year}^{-1}$$

$$\text{fraction stored} = \frac{7.26 \times 10^9}{7.96 \times 10^{12}} \approx 10^{-3}$$

17.16 (a) 0.1709 J s^{-1} (b) 0.3988 J s^{-1}
(c) 0.1709 watt (d) 0.3988 watt

17.17 2.18×10^{-19} J

17.18 3.94 tons

17.19 7.84×10^4 s

17.21 613 g

17.22 1.36×10^{17} molecules cm^{-2}

17.23 5.01×10^{-9} mol s^{-1}

17.24 0.050 J s^{-1}

17.25 $NO_2 + h\nu \longrightarrow NO_2^*$

$NO_2^* + NO_2 \longrightarrow 2NO + O_2$

Decreased concentration of NO_2 after long illumination makes decomposing collisions less likely. Also the reverse reaction becomes more important as the concentrations of the products build up.

17.28 24.4 tons

17.29 (a) 10^{14} (b) 4.53×10^{-9} g day^{-1}

17.30 504 nm

CHAPTER 18 Irreversible Processes in Solution

18.1 Ten cm^3 of water at 25° C is forced through 20 cm of 2 mm diameter capillary in 4 s. Calculate the pressure required and the Reynolds number.

SOLUTION:

$$P = \frac{8 V l \eta}{\pi r^4 t}$$

$$= \frac{(8)(10 \times 10^{-6} m^3)(0.2 m)(8.95 \times 10^{-4} Kg\, m^{-1} s^{-1})}{\pi (10^{-3} m)^4 (4 s)}$$

$$= 1.14 \times 10^3\ N\, m^{-2}$$

$$\text{Reynolds number} = \frac{d \bar{v} \rho}{\eta}$$

$$\bar{v} = \frac{\text{volume of liquid per unit time}}{\text{area of tube}}$$

$$= \frac{(10 \times 10^{-6} m^3)/(4 s)}{\pi (10^{-3} m)^2}$$

$$= 0.796\ m\ s^{-1}$$

Reynolds number

$$= \frac{(2 \times 10^{-3} m)(0.796\, m\, s^{-1})(10^3\, Kg\ m^{-3})}{(8.95 \times 10^{-4} Kg\, m^{-1} s^{-1})}$$

$$= 1780$$

18.2 A steel ball ($\rho = 7.86$ g cm^{-3}) 0.2 cm in diameter falls 10 cm through a viscous liquid ($\rho_o = 1.50$ g cm^{-3}) in 25 s. What is the viscosity at this temperature?

SOLUTION:

$$\eta = \frac{2r^2(\rho - \rho_o)g}{9\frac{dx}{dt}}$$

$$= \frac{2(1\times10^{-3}\,\mathrm{m})^2[(7.86-1.50)\times10^3\,\mathrm{Kg\,m^{-3}}](9.8\,\mathrm{m\,s^{-2}})}{9\left(\frac{0.10\,\mathrm{m}}{25\,\mathrm{s}}\right)}$$

$= 3.46$ Pa s

18.3 How long will it take a spherical air bubble 0.5 mm in diameter to rise 10 cm through water at 25° C?

SOLUTION:

$$\frac{\Delta X}{\Delta t} = \frac{2r^2(\rho - \rho_o)g}{9\eta}$$

$$\Delta t = \frac{9\eta\Delta X}{2r^2(\rho - \rho_o)g}$$

$$= \frac{9(8.95\times10^{-4}\,\mathrm{Kg\,m^{-1}\,s^{-1}})(-0.1\,\mathrm{m})}{2(0.25\times10^{-3}\,\mathrm{m})^2(-10^3\,\mathrm{Kg\,m^{-3}})(9.8\,\mathrm{m\,s^{-2}})}$$

$= 0.66$ s

18.5 A conductance cell was calibrated by filling it with a 0.02 mol L^{-1} solution of potassium chloride ($\kappa = 0.2768\ \Omega^{-1}$ m^{-1}) and measuring the resistance at 25°C, which was found to be 457.3 Ω. The cell was then filled with calcium chloride solution containing 0.555 gram of $CaCl_2$ per liter. The measured resistance was 1050 Ω. Calculate (a) the cell constant for the cell and (b) the conductivity of the $CaCl_2$ solution.

SOLUTION:

(a) From equation 18.8

$$K_{cell} = \kappa R$$

$$= (0.2768\ \Omega^{-1}\ m^{-1})(457.3\ \Omega)$$

$$= 126.6\ m^{-1}$$

(b)
$$\kappa = \frac{K_{cell}}{R}$$

$$= \frac{126.6\ m^{-1}}{1050\ \Omega} = 0.1206\ \Omega^{-1}\ m^{-1}$$

18.6 A moving boundary experiment is carried out with a 0.1 mol L^{-1} solution of hydrochloric acid at 25°C ($\kappa = 4.24\ \Omega^{-1}\ m^{-1}$). Sodium ions are caused to follow the hydrogen ions. Three milliamperes is passed through the tube of 0.3 cm^2 cross-sectional area, and it is observed that the boundary moves 3.08 cm in 1 hr. Calculate (a) the hydrogen-ion mobility, (b) the chloride-ion mobility, and (c) the electric field strength.

SOLUTION:

(a)
$$E = \frac{I}{A\kappa} = \frac{3 \times 10^{-3}\ A}{(0.3 \times 10^{-4}\ m^2)(4.24\ \Omega^{-1}\ m^{-1})}$$

$$= 23.58\ V\ m^{-1}$$

$$u = \frac{\Delta x/\Delta t}{E} = \frac{3.08 \times 10^{-2}\ m}{(60 \times 60\ s)(23.58\ V\ m^{-1})}$$

$$= 3.63 \times 10^{-7}\ m^2\ V^{-1}\ s^{-1}$$

(b) $\kappa = Fc\,(u_{H^+} + u_{Cl^-})$

$$u_{Cl^-} = \frac{\kappa}{Fc} - u_{H^+} = \frac{(4.24\ \Omega^{-1}\ m^{-1})}{(96485\ C\ mol^{-1})(0.1 \times 10^3\ mol^{-3})}$$

$$- 3.63 \times 10^{-7}\ m^2\ V^{-1}\ s^{-1}$$

$$= 7.64 \times 10^{-8}\ m^2\ V^{-1}\ s^{-1}$$

(c) See (a)

18.8 It is desired to use a conductance apparatus to measure the concentration of dilute solutions of sodium chloride. If the electrodes in the cell are each 1 cm^2 in area and are 0.2 cm apart, calculate the resistance which will be obtained for 1, 10, and 100 ppm NaCl a 25°C.

SOLUTION:

1 ppm = 1 g NaCl in 10^6 g H_2O

$$= \frac{(1 g)/(58.45\ g\ mol^{-1})}{1\ m^3}$$

$$= 1.71 \times 10^{-2}\ mol\ m^{-3}$$

Using electric mobilities at infinite dilution,

$\kappa = Fc\,(u_{Na} + u_{Cl^-})$

$$= (96{,}485\ C\ mol^{-1})(1.71 \times 10^{-2}\ mol\ m^{-3})[(5.192 + 7.913) \times 10^{-8}\ m^2\ V^{-1} s^{-1}]$$

$$= 2.16 \times 10^{-4}\ \Omega^{-1}\ m^{-1}$$

$$R = \frac{l}{\kappa A} = \frac{0.2 \times 10^{-2} \text{ m}}{(2.16 \times 10^{-4} \Omega^{-1} \text{m}^{-1})(0.01 \text{ m})^2}$$

$$= 9.25 \times 10^4 \Omega$$

For 10 ppm $R = 9.25 \times 10^3 \Omega$

For 100 ppm $R = 925 \Omega$

18.9 Calculate the conductivity at 25°C of a solution containing 0.001 mol L^{-1} hydrochloric acid and 0.005 mol L^{-1} sodium chloride. The limiting ionic mobilities at infinite dilution may be used to obtain a sufficiently good approximation.

SOLUTION:

$$\kappa = F \Sigma c_i u_i = F(c_{H^+} u_{H^+} + c_{Na^+} u_{Na^+} + c_{Cl^-} u_{Cl^-})$$

$$= (96,485 \text{ C mol}^{-1})[(1 \text{mol m}^{-3})(36.25) + (5 \text{mol m}^{-3})(5.192) + (6 \text{mol m}^{-3})$$

$$\times (7.91)] \times 10^{-8} \text{ m}^2 \text{ V}^{-1} \text{ s}^{-1}$$

$$= 0.1058 \ \Omega^{-1} \text{ m}^{-1}$$

18.10 Using Stokes' law calculate the effective radius of a nitrate ion from its mobility (74.0×10^{-9} m^2 V^{-1} s^{-1} at 25° C).

SOLUTION:

$$f_i = \frac{|Z_i| e}{u_i} = 6\pi \eta r$$

$$r = \frac{|Z_i| e}{u_i 6\pi \eta}$$

$$= \frac{1.602 \times 10^{-19} \text{ C}}{(74 \times 10^{-9} \text{ m}^2 \text{ V}^{-1} \text{s}^{-1})(6\pi)(8.95 \times 10^{-4} \text{ Kg m}^{-1} \text{ s}^{-1})}$$

$= 0.128$ nm

18.11 What is the self diffusion coefficient of Na^+ in water at 25°C ?

SOLUTION:

$$D = \frac{uRT}{|z|F}$$

$$= \frac{(5.192 \times 10^{-8}\ m^2\ V^{-1}\ s^{-1})(8.314\ J\ K^{-1} mol^{-1})(298.15\ K)}{9.6485 \times 10^4\ C\ mol^{-1}}$$

$$= 1.334 \times 10^{-9}\ m^2\ s^{-1}$$

18.12 Using a table of the probability integral, calculate enough points on a plot of C versus x (like Fig. 18.5c) to draw in the smooth curve for diffusion of 0.1 mol L^{-1} sucrose into water at 25°C after 4 hr. and 29.83 min. ($D = 5.23 \times 10^{-10}\ m^2\ s^{-1}$).

SOLUTION:

Normal probility function $= \frac{1}{\sqrt{2\pi}}\ e^{-y^2/2} =$ N.P.F.

Equation 18.32

$$c = \frac{c_0}{2}\left[1 + \frac{2}{\sqrt{\pi}}\int_0^{\frac{x}{2\sqrt{Dt}}} e^{-\beta^2}\, d\beta\right]$$

In order to get this equation in the form of the normal probiability function let $\beta^2 = t^2/2$. Thus $\beta = t/\sqrt{2}$ and $d\beta = dt/\sqrt{2}$.

$$C = \frac{c_0}{2}\left[1 + \frac{2}{\sqrt{2\pi}}\int_0^{\frac{x}{\sqrt{2Dt}}} e^{-t^2/2}\, dt\right]$$

$$= \frac{C_o}{2}\left[1 + \frac{1}{\sqrt{2\pi}}\int_{\frac{-x}{\sqrt{2Dt}}}^{\frac{x}{\sqrt{2Dt}}} e^{-t^2/2}\,dt\right]$$

This area is obtainable from tables of the normal probability integral (N.P.I.)

$$\frac{C_o}{2} = \left[1 \pm \text{Area of N.P.I.}\right]$$

The + sign is used when $x > 0$, and the − sign is used when $x < 0$.

At $x = 1$ cm the Gaussian parameter is

$$\frac{x}{\sqrt{2Dt}} = \frac{10^{-2}\ \text{m}}{\sqrt{2(5.23 \times 10^{-10}\,\text{m}^2\,\text{s}^{-1})(269.83 \times 60\,\text{s})}}$$

$$= 2.43$$

$\lvert x \rvert$ cm	$\frac{x}{\sqrt{2Dt}}$	Area N.P.I.	$c(x<0)$	$c(x>0)$
0	0	0	0.05 mol L^{-1}	0.05 mol L^{-1}
0.1	0.243	0.1922	0.040	0.060
0.2	0.486	0.3732	0.031	0.069
0.3	0.729	0.5340	0.023	0.077
0.4	0.972	0.6690	0.016	0.084
0.6	1.458	0.8556	0.007	0.093
0.8	1.944	0.9480	0.003	0.097
1.0	2.43	0.9850	0.002	0.098
1.2	2.916	0.9964	0.001	0.099

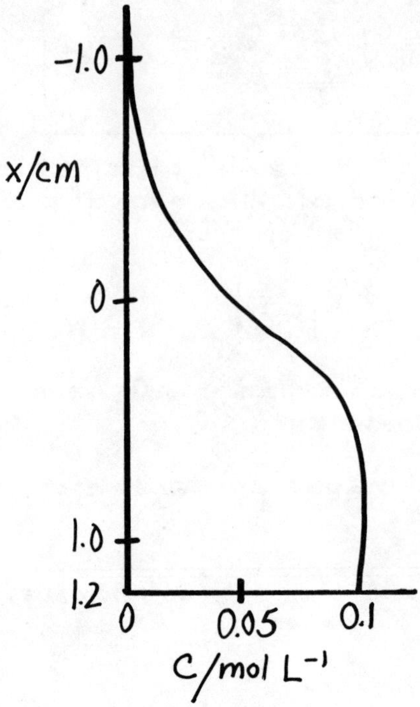

18.14 1.14×10^3 Pa m^{-2}
The Reynolds number is 447, and so the flow will be laminar.

18.15 95.7 min

18.16 3.01×10^{-3} cm s^{-1}

18.17 0.00141 Pa s

18.18 520 nm

18.19 (a) 126.5×10^{-4} Ω^{-1} m^2 mol^{-1} (b) 2.16×10^{-3} Ω^{-1} m^{-1}
(c) 7.36×10^4 Ω 1.09×10^{-3} A

18.20 $1.6 \times 10^{-3}\ \Omega^{-1}\ m^{-1}$

18.21 (a) $4.28\ \Omega^{-1}\ m^{-1}$ (b) 0.0768 m (c) $58.4\ V\ m^{-1}$

18.22 $3.16 \times 10^{-8}\ m^2\ V^{-1}\ s^{-1}$

18.23

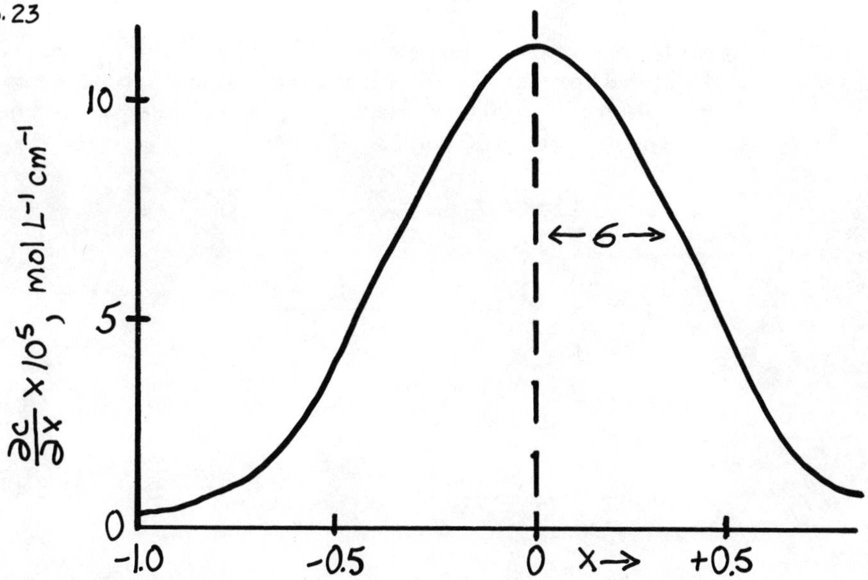

18.24 (a) 1.96 hr. (b) 56.4 hr

18.25 $6.43 \times 10^{-8}\ m^2\ V^{-1}\ s^{-1}$

18.28 (a) 117.7 min (b) 56.5 hr.

PART FOUR: STRUCTURE

CHAPTER 19. Crystal Structure and Solid State

19.1 Calculate the angles which the first-, second-, and third-order reflections are obtained from planes 500 pm apart, using X rays with a wavelength of 100 pm.

SOLUTION:

$$\sin\theta = \frac{\lambda}{2d_{hkl}} = \frac{\lambda}{2\,(d/n)}$$

$$\sin\theta_1 = \frac{100\text{ pm}}{2(500\text{ pm}/1)} \qquad \theta_1 = 5.74°$$

$$\sin\theta_2 = \frac{100\text{ pm}}{2(500\text{ pm}/2)} \qquad \theta_2 = 11.54°$$

$$\sin\theta_3 = \frac{100\text{ pm}}{2(500\text{ pm}/3)} \qquad \theta_3 = 17.46°$$

19.2 The crystal unit cell of magnesium oxide is a cube 420 pm on an edge. The structure is interpenetrating face centered. What is the density of crystalline MgO?

SOLUTION:

$$\frac{4(40.32\times10^{-3}\text{ kg mol}^{-1})}{(6.022\times10^{23}\text{ mol}^{-1})(4.20\times10^{-10}\text{ m})^3} = 3.615\times10^{3}\text{ kg m}^{-3}$$

$$= 3.615\text{ g cm}^{-3}$$

19.3 Tungsten forms body-centered cubic crystals. From the fact that the density of tungsten is 19.3 g cm^{-3} calculate (a) the length of the side of this unit cell and (b) d_{200}, d_{110} and d_{222}.

SOLUTION:

(a) $$19.3 \times 10^3 \text{ kg m}^{-3} = \frac{2(183.85 \times 10^{-3} \text{ kg mol}^{-1})}{(6.022045 \times 10^{23} \text{ mol}^{-1})a^3}$$

$$a = 316 \times 10^{-12} \text{ m} = 316 \text{ pm}$$

(b) $$d_{200} = \frac{a}{\sqrt{2^2+0+0}} = \frac{316 \text{ pm}}{2} = 158 \text{ pm}$$

$$d_{110} = \frac{a}{\sqrt{1+1+0}} = \frac{316 \text{ pm}}{\sqrt{2}} = 223 \text{ pm}$$

$$d_{222} = \frac{a}{\sqrt{4+4+4}} = \frac{316 \text{ pm}}{\sqrt{12}} = 91.2 \text{ pm}$$

19.4 Copper forms cubic crystals. When an x-ray powder pattern of crystalline copper is taken using x-rays from a copper target (the wavelength of the K_α line is 154.05 pm), reflections are found at $\theta = 21.65°$, 25.21°, 37.06°, 44.96°, 47.58°, and other larger angles. (a) What type of lattice is formed by copper? (b) What is the length of a side of the unit cell at this temperature? (c) What is the density of copper?

SOLUTION:

$$d_{hkl} = \frac{\lambda}{2 \sin \theta} = \frac{154.05 \text{ pm}}{2 \sin \theta} = \frac{a}{\sqrt{h^2+k^2+l^2}}$$

(a)

θ	d_{hkl}/pm	Ratio d_{hkl} to largest spacing
21.65°	208.77	1.0000
25.21°	180.84	0.8662
37.06°	127.81	0.6122
44.96°	109.00	0.5221
47.58°	104.34	0.4998

Ratios expected for cubic crystals ($d_{hkl} = \frac{a}{\sqrt{h^2+k^2+l^2}}$)

	Primitive	Body-centered	Face-centered
100	1.0000	—	—
110	0.7071	1.0000	—
111	0.5774	—	1.0000
200	0.5000	0.7071	0.8660
210	0.4472	—	—
211	X	0.5774	—
220	X	0.5000	0.6124
310	X	0.4472	—
311	X	—	0.5222
222	X	X	0.5000

— reflection absent X not needed for this problem

Thus copper forms face centered cubic crystals.

(b) $$d_{111} = \frac{a}{\sqrt{1+1+1}} = 208.77 \text{ pm}$$

$$a = 361.6 \text{ pm}$$

(c) $$d = \frac{4(63.546 \times 10^{-3} \text{ kg mol}^{-1})}{(6.02205 \times 10^{23} \text{mol}^{-1})(361.6 \times 10^{-12} \text{ m}^3)^3}$$

$$= 0.8927 \times 10^3 \text{ kg m}^{-3}$$

19.6 Cesium chloride, bromide, and iodide form interpenetrating simple cubic crystals rather than interpentrating face-centered cubic crystals like the other alkali halides. The length of the side of the unit cell of CsCl is 412.1 pm. (a) What is the density? (b) Calculate the ion radius of Cs^+ assuming that the ions touch along a diagonal through the unit cell and that the ion radius of Cl^- is 181 pm.

SOLUTION:

(a) $$d = \frac{(168.36 \times 10^{-3}\ \text{kg mol}^{-1})}{(6.022045 \times 10^{23}\ \text{mol}^{-1})(412.1 \times 10^{-12}\ \text{m})^3}$$

$$= 3.995 \times 10^3\ \text{kg m}^{-3}$$

(b) The diagonal plane through the cubic unit cell is as follows:

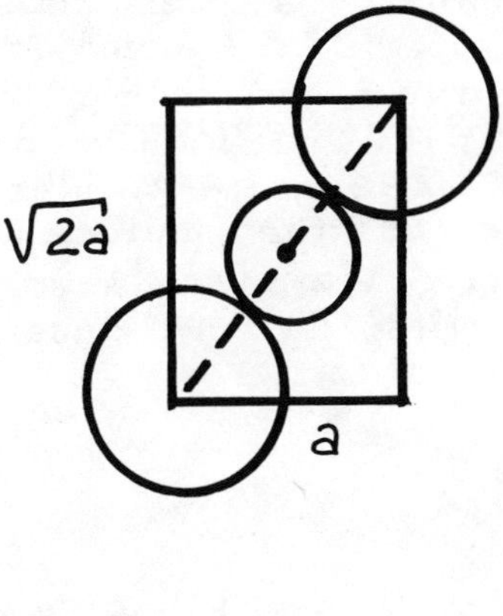

$$a^2 + (\sqrt{2}\,a)^2 = (2r_{Cs} + 2r_{Cl})^2$$

$$r_{Cs} = \frac{1}{2}\left[a^2 + (\sqrt{2}\,a)^2\right]^{1/2} - 2(181\ \text{pm})$$

$$= 176\ \text{pm}$$

19.7 The density of potassium chloride at 18° C is 1.9893 g cm^{-3}, and the length of a side of the unit cell is 629.082 pm, as determined by x-ray diffraction. Calculate the Avogadro constant using the values of the atomic masses given in the front cover.

SOLUTION:

$$1.9893 \times 10^3 \text{ kg m}^{-3} = \frac{4(74.551 \times 10^{-3} \text{ kg mol}^{-1})}{N_A (629.082 \times 10^{-12} \text{ m})^3}$$

$$N_A = 6.0213 \times 10^{23} \text{ mol}^{-1}$$

19.9 (a) Metallic iron at 20°C is studied by the Bragg method, in which the crystal is oriented so that a reflection is obtained from the planes parallel to the sides of the cubic crystal, then from planes cutting diagonally through opposite edges, and finally from planes cutting diagonally through opposite corners. Reflections are first obtained at θ = 11° 36′, 8° 3′, and 20° 26′, respectively. What type of cubic lattice does iron have at 20° C? (b) Metallic iron also forms cubic crystals at 1100°C, but the reflections determined as described in (a) occur at θ = 9° 8′, 12° 57′, and 7° 55′, respectively. What type of cubic lattice does iron have at 1100°C? (c) The density of iron at 20°C is 7.86 g cm^{-3}. What is the length of a side of the unit cell at 20 °C? (d) What is the wavelength of the X rays used? (e) What is the density of iron at 1100 °C?

SOLUTION:

(a) (b) The three orientations will give reflections from planes a00, bb0, and ccc, respectively, where a, b and c are small integers. The smallest suitable integers are: primitive lattice: a = b = c = 1; body-centered lattice: b = 1, a = c = 2 (100 and 111 reflections are missing); face-centered lattice: c = 1, a = b = 2.

$$\sin\theta = \frac{\lambda}{2d_{hkl}} = \frac{\lambda\sqrt{h^2+k^2+\ell^2}}{2a}$$

Since $\lambda/2a$ is constant for a particular experiment, the three lattice types can be distinguished simply on the basis of the relative magnitudes of the three θ's. Thus the crystal at 20°C is body-centered since the second angle is smaller than the first and the third is the largest of all (ratio of $\sin\theta$ is $2:\sqrt{2}:\sqrt{12}$). At 1100°C the face-centered form is found (ratio of $\sin\theta$ is $2:\sqrt{8}:\sqrt{3}$). (Note that a primitive lattice would give ratios of $\sin\theta$ of $1:\sqrt{2}:\sqrt{3}$).

(c) $d = 7.86\times10^3\ \text{kg m}^{-3} = \dfrac{2\,(55.847\times10^{-3}\ \text{kg mol}^{-1})}{(6.022045\times10^{23}\ \text{mol}^{-1})a^3}$

$$a = \left[\frac{(2)(55.847\times10^{-3}\ \text{kg mol}^{-1})}{(6.022045\times10^{23}\ \text{mol}^{-1})(7.86\times10^3\ \text{kg m}^{-3})}\right]^{1/3}$$

$= 286.8\times10^{-12}\ \text{m}$

$= 286.8\ \text{pm}$

(d) $\lambda = 2d_{hk\ell}\sin\theta = \dfrac{2a\sin\theta}{\sqrt{h^2+k^2+\ell^2}}$

$\dfrac{573.6\ \text{pm}}{2}\ \sin 11°36' = 57.7\ \text{pm}$

$\dfrac{573.6\ \text{pm}}{\sqrt{2}}\ \sin 8°3' = 56.8\ \text{pm}$

$\dfrac{573.6\ \text{pm}}{\sqrt{12}}\ \sin 20°26' = 57.8\ \text{pm}$

Average = 57.4 pm

(e) $d_{hk\ell} = \dfrac{\lambda}{2\sin\theta}$

$$d_{200} = \frac{57.4 \text{ pm}}{2 \sin 9° 8'} = 180.8 \text{ pm}; \; a = d_{200}\sqrt{4} = 361.6 \text{ pm}$$

$$d_{220} = \frac{57.4 \text{ pm}}{2 \sin 12° 57'} = 128.1 \text{ pm}; \; a = d_{220}\sqrt{8} = 362.3 \text{ pm}$$

$$d_{111} = \frac{57.4 \text{ pm}}{2 \sin 7° 55'} = 208.4 \text{ pm}; \; a = d_{111}\sqrt{3} = 360.9 \text{ pm}$$

Average $= 361.6$ pm

$$d = \frac{4(55.847 \times 10^{-3} \text{ kg mol}^{-1})}{(6.022045 \times 10^{23} \text{ mol}^{-1})(361.6 \times 10^{-12} \text{ m})^3}$$

$$= 7.846 \times 10^3 \text{ kg m}^{-3}$$

19.10 Insulin forms crystals of the orthorhombic type with unit-cell dimensions of 13.0 x 7.48 x 3.09 nm. If the density of the crystal is 1.315 g cm^{-3} and there are 6 insulin molecules per unit cell, what is the molar mass weight of the protein insulin?

SOLUTION:

$$d = 1.315 \times 10^3 \text{ kg m}^{-3} = \frac{6\,M}{(6.022045 \times 10^{23} \text{ mol}^{-1})(13.0 \times 7.48 \times 3.09 \times 10^{-27} \text{ m}^3)}$$

$$M = \frac{1}{6}(1.315 \times 10^3 \text{ kg m}^{-3})(6.022045 \times 10^{23} \text{ mol}^{-1})(13.0 \times 7.48 \times 3.09 \times 10^{-27} \text{ m}^3)$$

$$= 39.7 \text{ kg mol}^{-1}$$

$$= 39{,}700 \text{ g mol}^{-1}$$

19.12 Rutile (TiO_2) forms a primitive tetragonal lattice with $a = 459.4$ pm and $c = 296.2$ pm. There are two Ti atoms per unit cell, one at (000) and the other at $(\frac{1}{2}\frac{1}{2}\frac{1}{2})$. The four oxygen atoms are located at $\pm$ (uu0) and $\pm$ $(\frac{1}{2} + u, \frac{1}{2} - u, \frac{1}{2})$ with $u = 0.305$. What is the density of the crystal?

SOLUTION:

$$V = (459.4 \text{ pm})^2 (296.2 \text{ pm})$$

$$= 62.51 \times 10^{-30} \text{ m}^3$$

$$d = \frac{2(47.90 + 2 \times 15.99) \times 10^{-3} \text{ kg mol}^{-1}}{(6.022045 \times 10^{23} \text{ mol}^{-1})(62.51 \times 10^{-30} \text{ m}^3)}$$

$$= 4.245 \times 10^3 \text{ kg m}^{-3}$$

19.13 A solution of carbon in face-centered cubic iron has a density of 8.105 g cm^{-3} and a unit cell edge of 358.3 pm. Are the carbon atoms interstitial, or do they substitute for iron atoms in the lattice? What is the weight percent carbon?

SOLUTION:

$$d = \frac{4(55.847 \times 10^{-3} \text{ kg mol}^{-1})}{(6.022045 \times 10^{23} \text{ mol}^{-1})(358.3 \times 10^{-12} \text{ m})^3}$$

$$= 8.064 \times 10^3 \text{ kg m}^{-3}$$

Since the experimental density is greater, the carbon atoms must be interstitial.

8.105×10^3 kg m^{-3}

$$\frac{8.064 \times 10^3}{0.041 \times 10^3}$$

$$\frac{0.041 \times 10^3}{8.105 \times 10^3} \, 100 = 0.51\ \%\ \text{by weight}$$

19.15 If spherical molecules of 500 pm radius are packed in cubic close packing and body centered cubic, what are the lengths of the sides of the cubic unit cells in the two cases?

SOLUTION:

Cubic close packing (face centered cubic)

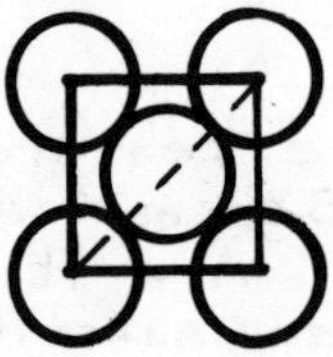

Length of diagonal = 2000 pm

Length of side of unit cell $= \frac{2000}{\sqrt{2}} = 1414.2$ pm

Body centered cubic

Consider a plane bisecting the cube and cutting diagonally through opposite faces.

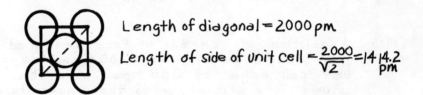

Length of diagonal = 2000 pm

$$2000^2 = a^2 + (\sqrt{2}a)^2 = 3a^2$$

$$a = \frac{2000^2}{\sqrt{3}} = 1154.7 \text{ pm}$$

19.16 Titanium forms hexagonal close-packed crystals. Given the atomic radius of 146 pm what are the unit cell dimensions and what is the density of the crystal?

SOLUTION:

$$a = b = 2\ (146\ \text{pm}) = 292\ \text{pm}$$

$$c = 2\sqrt{2}\ a / \sqrt{3} = 477\ \text{pm}$$

$$V = a^2 c\ (1 - \cos^2 \gamma)^{1/2} = a^2 c \sin \gamma$$

$$= (292\ \text{pm})^2\ (477\ \text{pm}) \sin 120°$$

$$= 35.22 \times 10^{-30}\ \text{m}^3$$

$$d = \frac{2\ (47.90 \times 10^{-3}\ \text{kg mol}^{-1})}{(6.022045 \times 10^{23}\ \text{mol}^{-1})(35.22 \times 10^{-30}\ \text{m}^3)}$$

$$= 4.517 \times 10^3\ \text{kg m}^{-3}$$

19.17 Metallic sodium forms a body centered cubic unit cell with $a = 424$ pm. What is the sodium atom radius?

SOLUTION:

Three atoms are in contact along the diagonal of the unit cell

$$(4R)^2 = (424\ \text{pm})^2 + (424\ \text{pm})^2 + (424\ \text{pm})^2$$

$$R = \frac{\sqrt{3}}{4} \; 424 \text{ pm} = 184 \text{ pm}$$

Alternatively,

$$\ell = a[(x_2 - x_1)^2 + (y_2 - y_1)^2 + (z_2 - z_1)^2]^{1/2}$$

$$= 424 \text{ pm} \left[\frac{1}{4} + \frac{1}{4} + \frac{1}{4}\right]^{1/2}$$

$$= \frac{\sqrt{3}}{4} = 424 \text{ pm} = 367 \text{ pm}$$

The sodium atom radius is half the distance.

$$\frac{367 \text{pm}}{2} = 184 \text{ pm}$$

19.18 The diamond has a face-centered cubic crystal lattice, and there are eight atoms in a unit cell. Its density is 3.51 g cm^{-3}. Calculate the first six angles at which reflections would be obtained using an X-ray beam of wavelength 71.2 pm.

SOLUTION:

$$d = 3.51 \times 10^3 \text{ kg m}^{-3} = \frac{8(12.011 \times 10^{-3} \text{ kg mol}^{-1})}{(6.022045 \times 10^{23} \text{ mol}^{-1}) a^3}$$

$$a = 357 \times 10^{-12} \text{ m} = 357 \text{ pm}$$

$$\theta = \sin^{-1} \frac{\lambda}{2a} \sqrt{h^2 + k^2 + \ell^2} = \sin^{-1} \frac{71.2}{2(357)} \sqrt{h^2 + k^2 + \ell^2}$$

For face-centered cubic the first six reflections are:

$hk\ell$	θ
111	9.95°
200	11.50°

220	16.38°
311	19.31°
222	20.21
400	23.51

19.20 At 550°C the conductivity of solid NaCl is $2 \times 10^{-4}\ \Omega^{-1}\ m^{-1}$. Since the sodium ions are smaller than the chloride ions (see table 19.1), they are responsible for most of the electric conductivity. What is the ionic mobility of Na^+ under these conditions?

SOLUTION:

Since there are 4 Na^+ per unit cell of a = 564 pm

$$N = \frac{4}{(564 \times 10^{-12}\ m)^3} = 2.23 \times 10^{28}\ m^{-3}$$

According to equation 19.30,

$$u = \kappa / Nq$$

$$= \frac{2 \times 10^{-4}\ \Omega^{-1}\ m^{-1}}{(2.23 \times 10^{28}\ m^{-3})(1.60 \times 10^{-19}\ C)}$$

$$= 5.61 \times 10^{-14}\ m^2\ V^{-1}\ s^{-1}$$

19.21 What fraction of the lattice sites of a crystal are vacant at 300 K if the energy required to move a atom from a lattice site in the crystal to a lattice site on the surface is 1 eV? At 1000 K?

SOLUTION:

At 300K

$$\frac{n}{N} = e^{-E_v/kT}$$

$$= e^{-\frac{(1V)(1.602\times10^{-19}C)}{(1.38\times10^{-23}JK^{-1})(300K)}}$$

$$= 1.56\times10^{-17}$$

At 1000K

$$\frac{n}{N} = e^{-\frac{(1V)(1.602\times10^{-19}C)}{(1.38\times10^{-23}JK^{-1}mol^{-1})(1000K)}}$$

$$= 9.08\times10^{-6}$$

19.22 A diamond has an electric conductivity of $5.1\times10^{-5}\ \Omega^{-1}\ m^{-1}$ at 25° C. Assuming that the mobilities of the carriers are independent of temperature calculate its conductivity at 35° C if the energy gap is 6 V.

SOLUTION:

$$\kappa = Ae^{-E_g/2RT}$$

$$A = \kappa e^{-E_g/2RT}$$

$$= (5.1\times10^{-5}\ \Omega^{-1}m^{-1})e^{\frac{(6V)(1.602\times10^{-19}C)}{2(1.381\times10^{-23}JK^{-1})(298K)}}$$

$$= 2.66\times10^{46}\ \Omega^{-1}\ m^{-1}$$

$$\kappa(35°) = 2.66\times10^{46}\Omega^{-1}m^{-1}\ e^{-\frac{(6V)(1.602\times10^{-19}C)}{2(1.381\times10^{-23}JK^{-1})(308K)}}$$

$$= 2.26\times10^{-3}\ \Omega^{-1}m^{-1}$$

19.23 $n = 7$

19.24 (a) 6 (b) 8 (c) 12

19.25

	Primitive	Body-centered	Face-centered
Nearest	6	8	12
Second nearest	12	6	6

19.26 152 pm

19.27 346 pm

19.29 (a) body-centered (b) 314.8 pm (c) 10.21 g cm^{-3}

19.30 (a) 628 pm (b) 2.00 g cm^{-3}

19.31 392.4 pm

19.32 74.69 g mol^{-1}

19.33 4

19.34 2826 kg m^{-3}

19.35 (a) 2, (b) 328, (c) 164, (d) 232, (e) 94.7 nm

19.36 2.703 g cm^{-3}

19.37 3.13 g cm^{-3}

19.38 (a) $\sqrt{8/3}\ a$ (b) $2\sqrt{2}\ r$ (c) $2r$

19.39 8.836×10^3 kg m^{-3}

19.40 0.414

19.41 2.33×10^3 kg m^{-3}

19.42 12

19.43 145.8 pm

19.44 490 $cm^2\ V^{-1}\ s^{-1}$

19.45 3.6×10^{-8} mole fraction

CHAPTER 20. Macromolecules

20.1 Calculate the sedimentation coefficient of tobacco mosaic virus from the fact that the boundry moves with a velocity of 0.454 cm hr^{-1} in an ultracentrifuge at a speed of 10,000 rpm at a distance of 6.5 cm from the axis of the centrifuge rotor.

SOLUTION:

$$s = \frac{\frac{dr}{dt}}{\omega^2 r} = \frac{\frac{0.454 \text{ cm hr}^{-1}}{3600 \text{ s hr}^{-1}}}{\left(\frac{10,000 \times 2\pi}{60 \text{ s}}\right)^2 6.5 \text{ cm}}$$

$$= 177 \times 10^{-13} \text{ s}$$

20.2 The sedimentation coefficient of myoglobin at 20°C is 2.06×10^{-13} s. What molar mass would it have if the molecules were spherical? Given: $\bar{v} = 0.749 \times 10^{-3}$ m^3 kg^{-1}, $\rho = 0.9982 \times 10^3$ kg m^{-3}, and $\eta = 0.001005$ Pa s.

SOLUTION:

$$s = \frac{M(1 - \bar{v}\rho)}{N_A f}$$

$$= \frac{M(1 - \bar{v}\rho)}{N_A 6\pi\eta} \left(\frac{4\pi N_A}{3M\bar{v}}\right)^{1/3}$$

$$M = \left(\frac{N_A 6\pi \eta s}{1 - \bar{v}\rho}\right)^{3/2} \left(\frac{3\bar{v}}{4\pi N_A}\right)^{1/2}$$

$$M = \left[\frac{(6.022 \times 10^{23} \text{ mol}^{-1}) 6\pi (0.001005)(2.06 \times 10^{-13})}{1 - (0.749 \times 10^{-3})(0.9982 \times 10^{3})} \right]^{3/2}$$

$$\times \left(\frac{3 \times 0.749 \times 10^{-3}}{4\pi\, 6.022 \times 10^{23}} \right)^{1/2}$$

$$= 15.5 \text{ kg mol}^{-1}$$

$$= 15,500 \text{ g mol}^{-1}$$

20.3 A sedimentation equilibrium experiment is to be carried out myoglobin ($M = 16,000$ g mol^{-1}) in an ultracentifuge operating at 15,000 revolutions per minute. The bottom of the cell is 6.93 cm from the axis of rotation and the meniscus is 6.67 cm from the axis of rotation. What ratio of concentrations is expected at 20° C if $\bar{v} = 0.75 \times 10^{-3}$ m^3 kg^{-1} and $\rho = 1.00 \times 10^3$ kg m^{-3}.

SOLUTION:

$$\ln \frac{c_2}{c_1} = \frac{M(1 - \bar{v}\rho)\,\omega^2 (r_2^{\,2} - r_1^{\,2})}{2RT}$$

$$= \frac{(16 \text{ kg mol}^{-1})(0.25)(2\pi 250 \text{ s}^{-1})^2 [(0.0693 \text{ m})^2 - (0.0667 \text{ m})^2]}{2\,(8.314 \text{ J K}^{-1}\text{mol}^{-1})\,(293.15 \text{ K})}$$

$$= 0.716$$

$$\frac{c_2}{c_1} = 2.05$$

20.4 The sedimentation and diffusion coefficients for hemoglobin corrected to 20°C in water are 4.41×10^{-13} s and 6.3×10^{-11} m^2 s^{-1}, respectively. If $\bar{v} = 0.749$ cm3 g^{-1} and $\rho_{H_2O} = 0.998$ g cm^{-3} at this temperature, calculate the molar mass of the protein. If there is 1 g atom of iron per 17,000 g of protein, how many atoms of iron are there per hemoglobin molecules?

SOLUTION:

$$M = \frac{RTS}{D(1-\bar{v}\rho)}$$

$$= \frac{(8.31\,J\,K^{-1}\,mol^{-1})(293\,K)(4.41\times10^{-13}\,s)}{(6.3\times10^{-11}\,m^2\,s^{-1})[1-(0.749\times10^{-3}\,m^3\,kg^{-1})(0.998\times10^3\,kg\,m^{-3})]}$$

$$= 67.6\ kg\ mol^{-1}$$

$$= 67,600\ g\ mol^{-1}$$

$$\frac{67,600\ g\ mol^{-1}}{17,000\ g\ g\text{-atom Fe}} = 4\ g\text{-}\ atom\ Fe\ mol^{-1}$$

$$= 4\ Fe\ per\ molecule$$

20.5 The diffusion coefficient for serum globulin at 20°C in a dilute aqueous salt solution is $4.0\times10^{-11}\ m^2\ s^{-1}$. If the molecules are assumed to be spherical, calculate their molar mass. Given: $\eta_{H_2O} =$ 0.001005 Pa s at 20°C and $\bar{v} = 0.75\ cm^3\ g^{-1}$ for the protein.

SOLUTION:

$$D = \frac{RT}{N_A 6\pi\eta}\left(\frac{4\pi N_A}{3M\bar{v}}\right)^{1/3}$$

$$M = \frac{4\pi N_A}{3\bar{v}} \left(\frac{RT}{N_A 6\pi \eta D}\right)^3$$

$$= \frac{4\pi(6.02 \times 10^{23} \text{mol}^{-1})}{3(0.75 \times 10^{-3} \text{m}^3 \text{kg}^{-1})}\left[\frac{(8.31 \text{J K}^{-1}\text{mol}^{-1})(293 \text{K})}{(6.02 \times 10^{23}\text{mol}^{-1}) 6\pi(1.005 \times 10^{-3} \text{J m}^{-3} \text{s})(4.0 \times 10^{-11} \text{m}^2 \text{s}^{-1})}\right]^3$$

$$= 511 \text{ kg mol}^{-1}$$

$$= 511{,}000 \text{ g mol}^{-1}$$

20.7 The diffusion coefficient of hemoglobin at 20° C is 6.9×10^{-11} m^2 s^{-1}. Assuming its molecules are spherical, what is the molar mass? Given: $\bar{v} = 0.749 \times 10^{-3}$ m^3 kg^{-1} and $\eta = 0.001005$ J m^{-3} s.

SOLUTION:

$$D = \frac{RT}{N_A 6\pi\eta} \left(\frac{4\pi N_A}{3M\bar{v}}\right)^{1/3}$$

$$M = \frac{4\pi N_A}{3\bar{v}} \left(\frac{RT}{DN_A 6\pi\eta}\right)^3$$

$$M = \frac{4\pi(6.022 \times 10^{23}\text{mol}^{-1})}{3(0.749 \times 10^{-3}\text{m}^3 \text{kg}^{-1})}\left[\frac{(8.314 \text{J K}^{-1}\text{mol}^{-1})(293 \text{K})}{(6.9 \times 10^{-11}\text{m}^2 \text{s}^{-1})(6.022 \times 10^{23}\text{mol}^{-1}) 6\pi(0.001005 \text{J m}^{-3} \text{s})}\right]^3$$

$$= 100 \text{ kg mol}^{-1} = 100{,}000 \text{ g mol}^{-1}$$

20.8 The protein human plasma albumin has a molar mass of 69,000 g mol^{-1}. Calculate the osmotic pressure of a solution of this protein containing 2 g per 100 cm^3 at 25°C in (a) torr and (b) millimeters of water. The experiment is carried out using a salt solution for solvent and a membrane permeable to salt.

SOLUTION:

$$\pi = \frac{cRT}{M}$$

$$= \frac{(20\ g\ L^{-1})(0.08205\ L\ atm\ K^{-1}\ mol^{-1})(298.15\ K)}{(69{,}000\ g\ mol^{-1})}$$

$$= 7.09 \times 10^{-3}\ atm$$

$$= (7.09 \times 10^{-3}\ atm)(760\ torr\ atm^{-1})$$

$$= 5.39\ torr$$

$$= 5.39\ mm\ Hg \quad \frac{13.534\ g\ cm^{-3}}{0.9970\ g\ cm^{-3}}$$

$$= 73.2\ mm\ of\ water$$

20.9 The following osmotic pressures were measured for solutions of a sample of polyisobutylene in benzene at 25 °C:

c, g/100 cm^3	0.500	1.00	1.50	2.00
x^2, g/cm^2	0.505	1.03	1.58	2.15

Calculate the number average moler mass from the value of π/c extrapolated to zero concentration of the polymer. [The pressures may be converted into atmospheres by dividing by (76 cm atm^{-1}) '3.53 g cm^{-3}) = 1028 g cm^{-2} atm^{-1}].

SOLUTION:

$$\frac{\Pi}{c} = \frac{RT}{M} + Bc$$

The ratio Π/c is plotted versus c and extrapolated to c = 0.

$c/g/100\ cm^3$	0.50	1.0	1.50	2.00
$\frac{\Pi\,(100)}{c\,1028}\Big/\frac{atm\ cm^3}{g}$	0.0982	0.1002	0.1025	0.1046

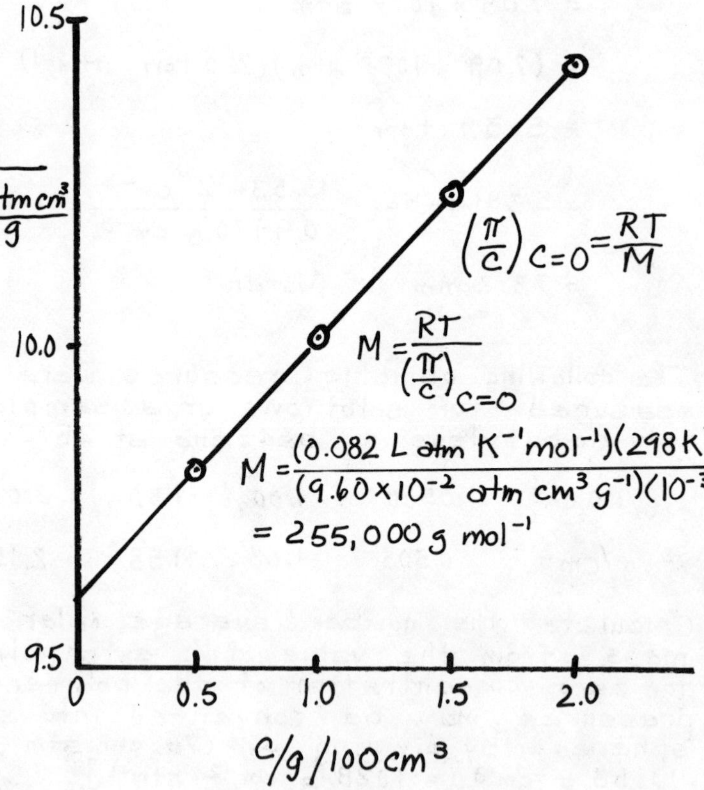

$$\frac{\frac{\Pi}{c}}{10^{-2}\frac{atm\ cm^3}{g}}$$

$$\left(\frac{\Pi}{c}\right)_{c=0} = \frac{RT}{M}$$

$$M = \frac{RT}{\left(\frac{\Pi}{c}\right)_{c=0}}$$

$$M = \frac{(0.082\ L\ atm\ K^{-1}mol^{-1})(298\ K)}{(9.60\times10^{-2}\ atm\ cm^3\ g^{-1})(10^{-3}\ L\ cm^{-3})}$$

$$= 255{,}000\ g\ mol^{-1}$$

20.10 A beam of sodium D light (589 nm) is passed through 100 cm of an aqueous solution of sucrose containing 10 g sucrose per 100 cm^3. Calculate I/I_o, where I_o is the intensity that would have been obtained with pure water, given that $M = 342.30$ g mol^{-1} and $dn/dc = 0.15$ $g^{-1}cm^3$ for sucrose. The refractive index of water at 20°C is 1.333 for the sodium D line.

SOLUTION:

$$\tau = \frac{32\pi^3 n_o^2 (dn/dc)^2 Mc}{3N_A \lambda^4}$$

$$= \frac{32\pi^3(1.333)^2(0.15\ g^{-1}cm^3)^2(342.30\ g\ mol^{-1})(0.1 g\ cm^{-3})}{3(6.02\times10^{23}\ mol^{-1})(5.89\times10^{-5}\ cm)^4}$$

$$= 6.25\times10^{-5}\ cm^{-1}$$

$$\frac{I}{I_o} = e^{-\tau x}$$

$$= e^{-(6.25\times10^{-5}\ cm^{-1})(100\ cm)}$$

$$= 0.9938$$

20.11 The relative viscosities of a series of solutions of a sample of polystyrene in toluene were determined with an Ostwald viscometer at 25°C.

$c/g/100\ cm^3$	0.249	0.499	0.999	1.998
η/η_o	1.355	1.782	2.879	6.090

The ratio η_{sp}/c is plotted against c and extrapolated to zero concentration to obtain the intrinsic viscosity. If the constants in equation 20.19 are $K = 3.7\times10^{-4}$

and $a = 0.62$ for this polymer, calculate the molar mass.

SOLUTION:

$c/g/100\ cm^3$	0.249	0.499	0.999	1.998
$\frac{\eta/\eta_0 - 1}{c}$	1.426	1.567	1.881	2.548

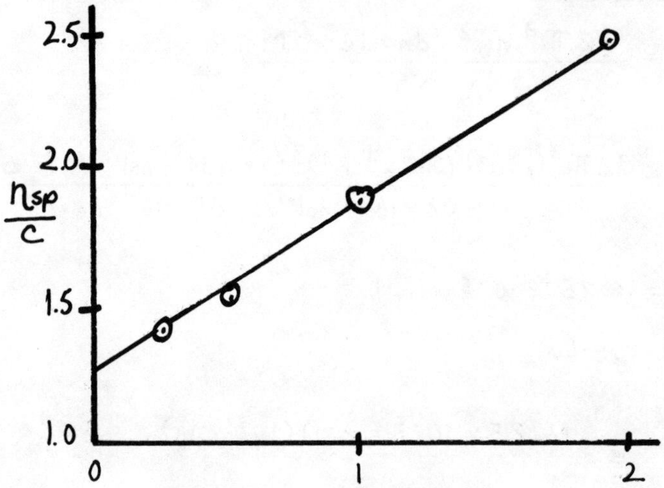

$[\eta] = 1.26 = (3.7 \times 10^{-4})\ M^{0.62}$

$M = 497{,}800\ g\ mol^{-1}$ which should be rounded off to $500{,}000\ g\ mol^{-1}$.

20.12 At 34 °C the intrinsic viscosity of a sample of polystyrene in toluene is $0.84\ dL\ g^{-1}$. The empirical relation between the intrinsic viscosity of polystyrene in toluene and molar mass is

$$[\eta] = 1.15 \times 10^{-4} M^{0.72}$$

when $[\eta]$ is expressed in dL g^{-1}.
What is the molar mass of this sample?

SOLUTION:

$$\ln [\eta] = \ln 1.15 \times 10^{-4} + 0.72 \ln M$$

$$\ln M = \frac{\ln [\eta] - \ln 1.15 \times 10^{-4}}{0.72}$$

$$M = e^{\frac{\ln [\eta] - \ln 1.15 \times 10^{-4}}{0.72}}$$

$$= e^{\frac{\ln 0.84 - \ln 1.15 \times 10^{-4}}{0.72}}$$

$$= 296{,}000 \text{ g mol}^{-1}$$

20.15 For a condensation polymerization of a hydroxyacid in which 99% of the acid groups are used up, calculate (a) the average number of monomer units in the polymer molecules, (b) the probability that a given molecule will have the number of residues given by this value, and (c) the weight fraction having the particular number of monomer units.

SOLUTION:

(a) The number-average degree of polymerization is given by

$$\bar{x}_n = \frac{1}{1-p} = \frac{1}{1-0.99} = 100$$

(b) The probility that a given molecule will have 100 residues is given by

$$\Pi_i = p^{i-1}(1-p)$$

$$\Pi_{100} = 0.99^{100-1}(1-0.99) = 3.70 \times 10^{-3}$$

(c) The weight fraction having this number of residues is given by

$$W_i = i p^{i-1}(1-p)^2$$

$$W_{100} = (100)(0.99)^{100-1}(1-0.99)^2$$

$$= 3.70 \times 10^{-3}$$

20.16 In the condensation polymerization of a hydroxyacid with a residue weight of 200 it is found that 99% of the acid groups are used up. Calculate (a) the number average molar mass and (b) the mass average molar mass.

SOLUTION:

(a) $$\overline{X}_n = \frac{1}{1-P} = \frac{1}{1-0.99} = 100$$

$$M_n = (100)(200\ \text{g mol}^{-1}) = 20{,}000\ \text{g mol}^{-1}$$

(b) $$\frac{M_m}{M_n} = 1 + p = 1.99$$

$$M_m = 39{,}800\ \text{g mol}^{-1}$$

20.17 When rubber is allowed to contract, the mechanical work obtained is equal to the force times the displacement in the direction of the force and is given by $dw = -f\,dL$, where f is force and L is the length of the piece of rubber. If the contraction is carried out reversibly, the first law may be written

$$dV = \left(\frac{\partial V}{\partial T}\right)_L dT + \left(\frac{\partial V}{\partial L}\right)_T dL = T\,dS + f\,dL$$

if pressure-volume work is neglected. (a) Show that

$$\left(\frac{\partial S}{\partial T}\right)_L = \frac{1}{T}\left(\frac{\partial V}{\partial T}\right)_L \qquad \left(\frac{\partial S}{\partial L}\right)_T = \frac{1}{T}\left[\left(\frac{\partial V}{\partial L}\right)_T - f\right]$$

(b) By using the fact that the order of differentiation used to obtain $\partial^2 S/\partial L\,\partial T$ is immaterial, show that

$$\left(\frac{\partial V}{\partial L}\right)_T = f - T\left(\frac{\partial f}{\partial T}\right)_L$$

SOLUTION:

(a) $$dS = \frac{1}{T}dU - \frac{f}{T}dL + \frac{dU}{T}$$

$$= \frac{1}{T}\left(\frac{\partial U}{\partial T}\right)_L dT + \frac{1}{T}\left[\left(\frac{\partial U}{\partial L}\right)_T - f\right] dL$$

$$= \left(\frac{\partial S}{\partial T}\right)_L dT + \left(\frac{\partial S}{\partial L}\right)_T dL$$

$$\left(\frac{\partial S}{\partial T}\right)_L = \frac{1}{T}\left(\frac{\partial U}{\partial T}\right)_L \qquad \left(\frac{\partial S}{\partial L}\right)_T = \frac{1}{T}\left[\left(\frac{\partial U}{\partial L}\right)_T - f\right]$$

(b) $$\frac{\partial^2 S}{\partial T\,\partial L} = \frac{1}{T}\left(\frac{\partial^2 U}{\partial T\,\partial L}\right) = -\frac{1}{T^2}\left[\left(\frac{\partial U}{\partial L}\right)_T - f\right] + \frac{1}{T}\left[\frac{\partial^2 U}{\partial T\,\partial L} - \frac{\partial f}{\partial T}\right]$$

$$\left(\frac{\partial U}{\partial L}\right)_T = f - T\left(\frac{\partial f}{\partial T}\right)_L$$

20.18 In polyethene $H(CH_2-CH_2)_nH$ the bond length b is about 0.15 nm. What is the root-mean-square end-to-end distance for a molecule with a molar mass of 10^5 g mol^{-1}? Taking into account the fact that carbon forms tetrahedral bonds, what is $\langle L^2\rangle^{1/2}$?

SOLUTION:

$$N = \frac{10^5 \text{ g mol}^{-1}}{14 \text{ g mol}^{-1}}$$

$$\langle L^2 \rangle^{1/2} = N^{1/2} b = \left(\frac{10^5}{14}\right)^{1/2} 0.15 \text{ nm}$$

$$= 12.7 \text{ nm}$$

$$\langle L^2 \rangle^{1/2} = N^{1/2} b \left(\frac{1 + \cos\theta}{1 - \cos\theta}\right)^{1/2}$$

$$= 12.7 \text{ nm} \left(\frac{1 + \cos 71°}{1 - \cos 71°}\right)^{1/2}$$

$$= 17.8 \text{ nm}$$

20.20 0.33 cm

20.21 64,000 g mol

20.22 2.37×10^{-14} s

20.23 7.09×10^{-11} m^2 s^{-1}

20.24 27.6×10^6 g mol^{-1}

20.25 122,000 g mol^{-1}

20.26 13.6 torr

20.27 19,500 g mol^{-1} 0.9980

20.28 8.39×10^6 g mol^{-1}

20.29 638,000 g mol^{-1}

20.30 (a) 36,800 g mol^{-1} (b) 48,500 g mol^{-1}

20.31 (a) 20 (b) 0.0189 (c) 0.0189

20.32 $20 M_0$, $39.0\ M_0$.

20.33 (a) 1100 nm (b) 13 nm (c) 18.4 nm

20.34 The maximum in the plot is at r = 10 nm.